BeginneR

Introductory Statistics Using R

Darrin Thomas

BeginneR
Introductory Statistics Using R

Darrin Thomas

SuJinSoLa
Saraburi, Thailand

Layout: Darrin Thomas
Photo Researcher: Darrin Thomas
Cover Design: Darrin Thomas

Copyright © 2018 ERT Group, publishing as SuJinSoLa, Saraburi, Thailand. All rights reserved.

SuJinSoLa

ISBN-13:978-1719554299
ISBN-10:1719554293

About the Author

Darrin Thomas, PhD, grew up in Sacramento, California and has over ten years of experience as a teacher and lecturer from Kindergarten to graduate school. He completed his bachelor and master degree in saxophone performance at California State University Sacramento. After working as a substitute teacher, he completed a credential in teaching at Pacific Union College. He then worked as a music teacher before moving to Thailand to work as a lecturer in Education/Psychology Department at Asia-Pacific International University (APIU). While overseas, Dr. Darrin completed his master degree in education at APIU. He then moved to the Philippines and completed his doctoral degree in education at Adventist International Institute of Advanced Studies. Currently, Dr. Darrin is a lecturer at Asia-Pacific International University. His enthusiasm for machine learning has led to works involving many different algorithms applied in an educational context. His blog is at `https://educationalresearchtechniques.com`

Dedication

To my Wife and Children

Contents

Preface   **vii**

1   What is Statistics?   **1**
     Introduction 2
     Types of Statistics 3
     Types of Variables 5
     More on Variables 8
     Statistical Notation 11
         Example 1.1 11
         Example 1.2 12
         Example 1.3 12
         Example 1.4 13
     Conclusion 13
     Points to Remember 13
     Exercises 14
         1.1 14
         1.2 14
         1.3 15

2   How do You Use R?   **17**
     Introduction 18
     Installing R 18
     Installing RStudio 20
     R Basics 21
     Using R 22
     Functions 23
     Loading Default Data 25
     Loading Your Own Data 26
     How to Think Like a Coder 26
     Conclusion 26
     Points to Remember 26

R Coding Used	27
Exercises	27
2.1	27
2.2	27
2.3	28
2.4	28

3 How do You Visualize Numbers — 29

Introduction	30
Frequency Table with Categorical Variable	30
Example 3.0 Table for Categorical Variables	31
Example 3.1: Using R to Create a Frequency Table	32
Frequency Table with Continuous variable	34
Example 3.2	35
Example 3.3: Using R for Frequency Table with Continuous Variable	38
Stem and Leaf Plot	42
Histogram	43
Bar Graph	44
Pie Graph	45
Scatter Plot	46
Exporting Figures to Word	47
Conclusion	50
Points to Remember	50
R Code Used	50
Exercises	51

4 What are Measures of Central Tendency — 53

Introduction	54
Mean	54
Example 4.1	55
Find the Mean Using R	55
Median	55
Find Median Using R	56
Mode	56
Conclusion	56
Points to Remember	56
R Code Used	57
Exercises	57
4.1	57

5 What are Measures of Dispersion?   59

     Introduction 60
     Range 60
         Example 5.1: Find Range Using R 60
     Variance & Standard Deviation 61
         Example 5.1 67
         Find Standard Deviation Using R 69
     Quartiles 69
         Example 5.2 70
         Finding Quartiles Using R 71
     Box Plots 71
         Make Box Plot Using R 73
     Kurtosis 74
         Find Kurtosis Using R 76
     Skewness 76
         Find Skewness Using R 77
     Conclusion 77
     Points to Remember 77
     R Code Used 78
     Exercises 78

6 What is Probability?   79

     Introduction 80
     Definition and Example 80
         Continuous Probability 82
     Conclusion 84

7 What is Normal Distribution?   85

     Introduction 86
     Understanding Normal Distribution 86
     Standard Normal Distribution 92
         Example 7.1 93
         Example 7.2 94
         Example 7.3 95
         Example 7.4 96
     Putting it All Together 97
     Conclusion 99
     Points to Remember 99
     R Code Used 99
     Exercises 99

8 What are Confidence Intervals? — 101
- Introduction 102
- Understanding Confidence Intervals 102
  - Example 8.1 108
  - Finding Confidence Intervals Using R 108
- Confidence Interval for Proportions-Categorical 109
  - Example 8.2 109
  - Finding Proportion Confidence Intervals Using R 110
- Conclusion 110
- Points to Remember 110
- R Code Used 110
- Exercises 110

9 What is a Hypothesis Testing? — 113
- Introduction 114
  - Example 9.1 114
- One Sample t-test 119
  - Example 9.2 120
  - Example 9.3 121
  - One Sample t-test Using R 123
- t-test for proportion: Categorical Data 124
  - Example 9.4 124
  - Example 9.5 126
  - Test of Proportions Using R 128
- Conclusion 129
- Points to Remember 129
- R Code Used 129
- Exercises 129

10 What is Two Sample Hypothesis Testing? — 131
- Introduction 132
- T-Test for Two Means 132
  - Example 10.1: Two-Tailed Test 134
  - Example 10.2: One-Tailed Test 135
  - Two Sample t-test Using R 136
- Paired T-Test 136
  - Example 10.3 138
  - Paired t-test Using R 139
- Two-Sample Test of Proportions 140
  - Example 10.4 141
  - Two-Sample Test of Proportion Using R 141

Conclusion	142
Points to Remember	142
Exercises	142

11 What is Analysis of Variance (ANOVA)?   145

Introduction	146
The Process	146
Example 11.1	148
Conclusion	149
Points to Remember	149
R Code Used	149
Exercises	149

12 What is Correlation and Regression?   151

Introduction	152
Scatter Plots and Correlation	152
Example 12.1	155
Example 12.2	156
Simple Linear Regression	157
Example 12.3	160
Multiple Regression	163
Example 12.4	163
Conclusion	165
Points to Remember	165
R Code Used	165
Exercises	165

13 What is Chi-Square?   167

Introduction	168
Goodness of Fit	168
Example 13.1	169
Find Goodness of Fit Using R	170
Test of Independence	170
Example 13.2	170
Conclusion	171
Points to Remember	171
R Code Used	171
Exercises	171
Mean and Standard Deviation	172
Percentages	172
T-Test	172

ANOVA 172
Correlation 172
Regression 173
Chi-Square 173

Preface

The purpose of this text is to provide a practical explanation of statistics with an application of its use in the R computer language. There is no detailed explanation of the foundational theoretical principles of algorithms and abstract data structures. This is a book for practitioners who want to get things done rather than dwell on the "why" of theories. Theories are important but not necessarily for everyone.

With the emergence of Data Science it is becoming harder and harder to ignore the need of understanding how to analyze and interpret data. In addition, with the growth of Big Data it is also becoming impractical to rely on point and click software such as Microsoft Excel. Practitioners of data analyst must understand how to analyze data as well as how to do this through use of simple coding. This text attempts to serve this need.

The concepts covered in this text are what are expected to be addressed in a introduction to statistics class. Topics include mean, standard deviation, probability, normal distribution, t-test, ANOVA, and correlation/regression. How to perform all of these different analyses in R is explain with example and exercises at the end of the chapters when needed.

You do not need any prior knowledge of computer coding or statistics to understand this book. This text focuses on what a practitioner needs to know rather than an expert. In addition, the coding is the bare minimum to complete the various task and no one will be an expert after completing this introductory text.

Hopefully, this book will provide you with the tools you need to perform statistical analysis for whatever purposes you may have.

Chapter 1

What is Statistics?

In this chapter, you will learn the following...

- What exactly descriptive and inferential statistics are.
- The different types of measurement and variables used in statistical research.
- How statistical notation works.

Key Terms.

Descriptive statistics	Inferential statistics	Categorical variable	Continuous variable
Nominal variable	Ordinal variable	Interval variable	Ratio variable

Introduction

One way to define statistics is as the practice of collecting and analyzing quantitative data. By quantitative we mean numbers. Statistics are used in a variety of ways in everyday life. For example, if you are reading something on the Internet, perhaps the author says the following...

The average life expectancy from birth in the world today is 71.5 years.

The key word in that sentence above is the word **average**. This is a term you are probably already familiar with and we will discuss this more in a future chapter. Below is another example,

The median price of a home in the US is $189,000.

Here the important term is the word **median**. If you don't know what this is you will learn later in this text. The point for now is that statistics is used everywhere in daily life. In addition, statistics is frequently used to answer questions about the world around us as well. Below are some practical questions that can be answered with an understanding of statistics.

- What is the average height of a student?

- What is the most common major studied on-campus?

- What is average price of a meal at lunch time?

- Do men or women have a higher GPA?

- How does sleep affects exam performance?

As you can see, statistics is a tool that helps us understand what we have observed through the use of numbers. Knowledge of the use of numbers as well as their meaning or interpretation can provide useful insights. Below are some more complex questions that statistics can deal with as well.

- How to spend money based on department size?

- Who is more likely to fail a course?

- What is a student's predicted test score based on class attendance and online activity?

Generally, what all statistics does is take numerical data from a sample or population and summarize it in different ways based on the questions that the you are seeking answers too. This aggregation (combining) of the data helps you to understand what is happening within your sample.

Types of Statistics

There are two commonly used types of statistics in research. These are descriptive and inferential statistics. **Descriptive statistics** are used for describing a group. An example of a descriptive statistic is the mean or average. The average tells you in one number what the typical value would be for a given group. For example, if the average height of men is 182cm you would expect the typical male to be this height. Remembering this one number is much easier than trying to remember all the individual heights of every male, especially if you measured many males. In addition, having one number is easier to explain and interpret.

To understand **inferential statistics** it is important that we understand what the word inferential means. Inferential comes from the word inference. An inference is a decision, opinion, or judgment that you make based on the information that you have at that moment. In other words, an inference allows you to make a decision about what you do not know based on what you do know.

Inferential statistics involves a sample, population, some form of sampling, and a question you want an answer too. For example, let us say we have 1,000 students on-campus. We need to find the average height of a student on-campus. So far we know our population which is 1,000 students and we know our question which is "what's the average height?" We will list this in the space below.

- Population = 1,000 students

- Question = What is the average height?

We do not have time to measure the height of every single student in the population. Therefore, we only take the height of some students. In other words, we draw a sample, from the population. A sample is a portion of the population. Let us say we sample 50 students. Below is an update of the information we have.

- Population = 1,000 students
- Question = What is the average height?
- Sample size = 50 students

One more thing we have to decide is how to draw the sample. For simplicity, let us say we randomly pick students to participate. By random, we mean that everyone has the same chance to be in our sample and we select 50 students. This is our final piece of information to solve this problem.

- Population = 1,000 students
- Question = What is the average height?
- Sample size = 50 students
- Sampling = Random

Figure 1.1 is a visual of how the sample is pulled from the population.

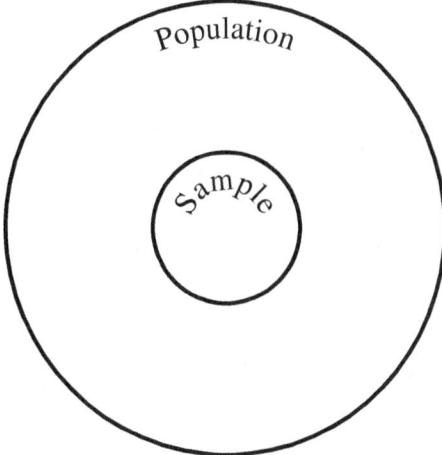

Figure 1.1: Sample vs Population

Once we measure the 50 people, we then calculate the average height. Let us say the average height in the sample of 50 students is 182cm. We can therefore make an inference that the average height of everybody on-campus is about 182cm. In other words, based on the sample of 50 university students we make the inference or conclusion that what is true in the sample is true in the population of the university. What we did here is we took what we knew, which was the sample height of the students, and made an inference that everybody was about the same height as the height we recorded in the sample. Now real inferential statistics is slightly more complicated but we will discuss this later.

There is a technical difference between descriptive and inferential statistics. Descriptive statistics describes populations while inferential statistics describes samples. The mean or any other measured value of descriptive study is a parameter but in inferential statistics, the same mean or measured value is a statistic. This is extremely confusing and is not critical at this point. However, for most of this book we will be using inferential statistic terminology.

Types of Variables

A variable is a feature within a sample or population that changes or has variation to it. Sometimes instead of using the term variable we talk about measurement instead. What this means is the type of variable you are using is also a type of measurement. This will make more sense when you conduct research rather than read about it.

There are two broad classes of variables and they are categorical and continuous. Within each class, there are two types of variables. For categorical, they are nominal and ordinal and for continuous they are interval and ratio. Figure 1.2 provides a visual of this.

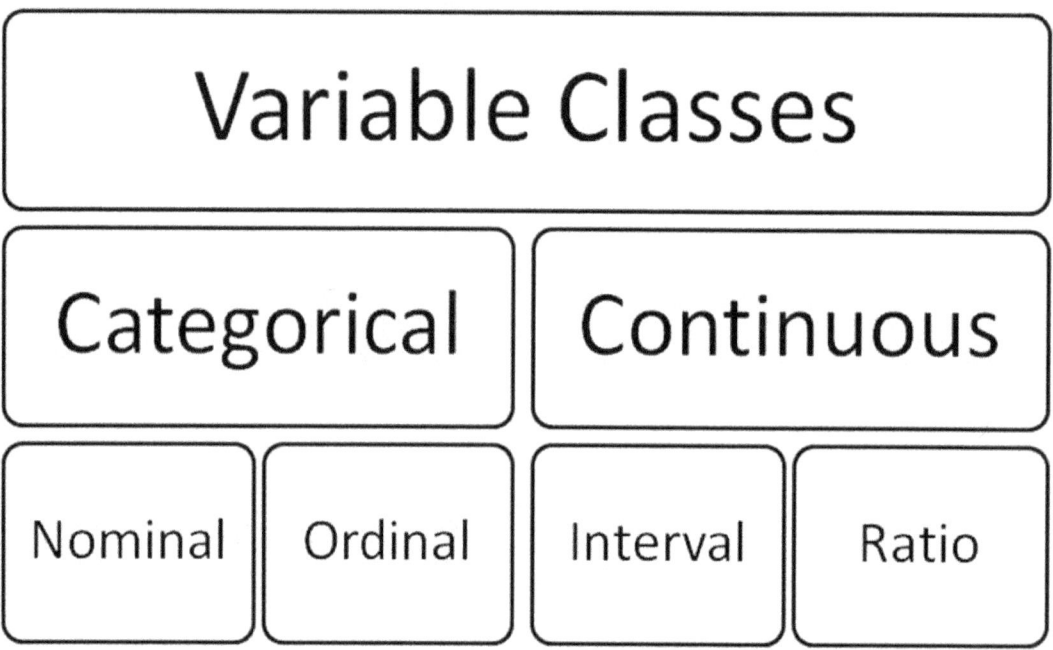

Figure 1.2: Variable Types

As you can see in Figure 1.2, under categorical variables, we have nominal and ordinal. A nominal variable differentiates by class. Examples of nominal variables include gender, type of fruit, country of origin, etc. When something is measured from the perspective of nominal data, it is placed in a category. In other words, there is no fuzziness to the measurement. Someone is male or female, Christian or Buddhist but not both. This is the weakest way to measure something because there is no way to tell how different two categories are as this is not measured. For example, if our categorical variable is gender there is no way to know by how much a man is different from a woman and vice versa because these are qualitative traits and not really quantitative. You can count how many men and women there are but you can not determine how different they are from each other. Figure 1.3 is a visual example of the categorical variable of gender.

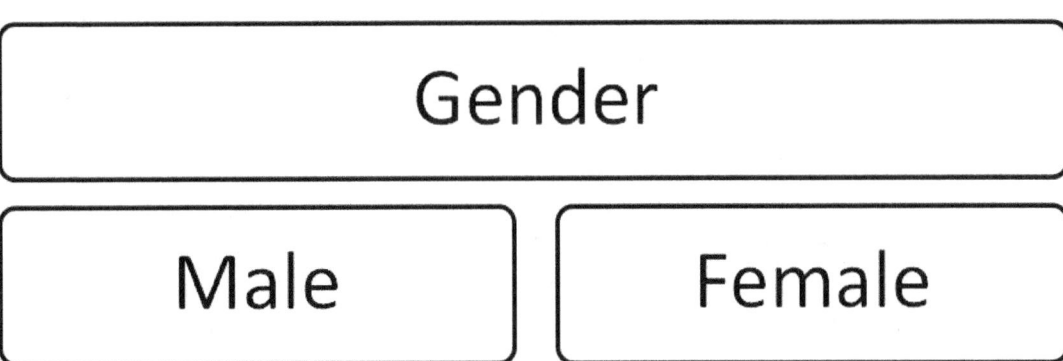

Figure 1.3: Nominal Variable

The next level of measurement is ordinal. Ordinal has all the characteristics of nominal variables with the addition of ranking. Ranking indicates that a category has a higher or lower value than another category. For example, faculty rank at a university, an assistant professor has a lower rank then an associate professor. In this example, we have two categories, just as in a nominal variable, but we also have a difference in rank in that an associate professor is above or of a higher rank than an assistant professor is. Now we know there is a difference in term of more or less but the amount of distance is not clearly defined. In other words, an associate professor is higher than an assistant professor is but the amount of difference is still mysterious. Figure 1.4 is a visual of the ordinal variable of academic rank.

Academic Rank

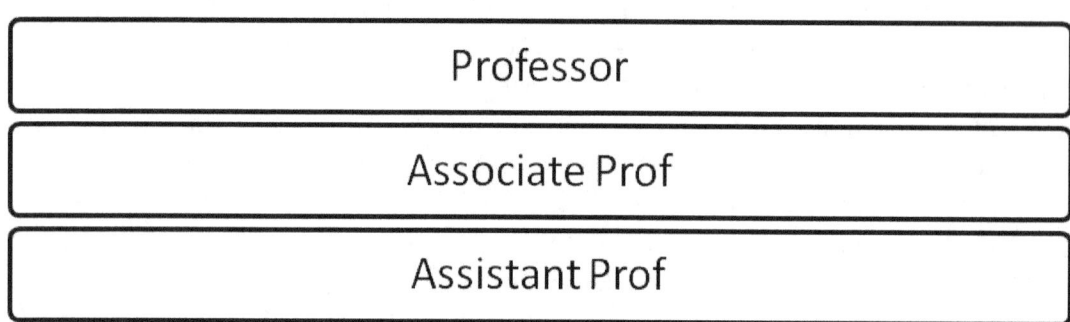

Figure 1.4: Ordinal Variable

We will now look at continuous variables. Continuous variables are fuzzy in that there are normally an infinite number of values a specific example can take. Under continuous variables we have interval and ratio variables. Interval variables has all the characteristics of ordinal (and thus nominal) with the addition that the amount of difference is indicate. An example, is temperature. The difference between 10 and 20 degrees Celsius is clear (10 degrees). There is no mystery that 20 degrees is warmer or that 10 degrees is cooler as we can clearly see what the difference is. Furthermore, thanks to decimals the number of potential values are endless. For example, we can have such measurements as 20.34, 20.6789, 20.1, etc. Lastly, interval measurement can include negative numbers such as -15 Celsius. Figure 1.5 is a visual of a interval variable of temperature.

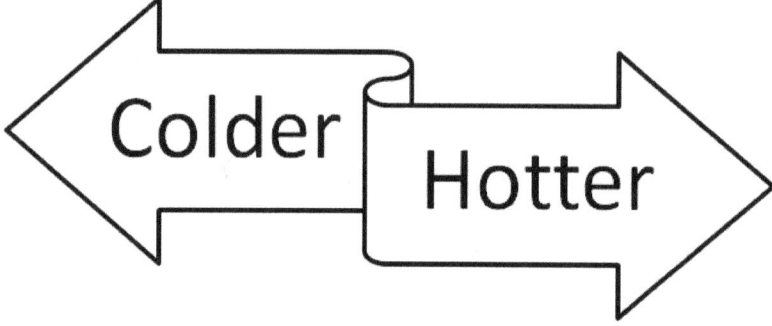

Figure 1.5: Interval Variable

Ratio measurement has all the characteristics of interval, ordinal, and nominal but it includes an absolute zero, which means that there can be no negative values. Examples of variables that are normally measured using ratio include salary, height, and weight. We can all agree that a person cannot have a negative weight or height. Table 1.1 is a summary of the different types of variables.

Measurement Type	Characteristics
Nominal	Differentiates class
Ordinal	Nominal traits plus rank
Interval	Ordinal traits plus indicates the amount of difference
Ratio	Interval traits plus absolute zero

Table 1.1: Types of Variables

Which type of measurement or variable to use depends less on what you are measuring and more on your research questions. For example, if I am measuring student GPA I can measure it categorical

What is your GPA? __ 0-1 __ 1-2 __ 2-3 __ 3-4

        or continuously

Write your GPA on the line ____

Both of these methods measure the same variable different ways. The method you use depends on what exactly it is you want to know.

More on Variables

So far we have discussed how variables can be categorical or continuous and we have looked at how there are different ways in which variables can be measured. We now will turn our attention to understanding the relationship between variables rather than focusing on the characteristics of a variable.

Variables can also be dependent or independent. Dependent variables are influenced by independent variables. As such, independent variables influence the dependent variable. What makes this more confusing is that either categorical or continuous variables can be independent or dependent. Again, it all depends on the type of measurement and question that is asked. Below are some examples of the various potential combinations of relationships between variables. Figure 1.6 is an example of a continuous independent and continuous dependent variable. The question is above the figure.

- Question-Is there a relationship between the amount of sleep students receive and their performance on a quiz?

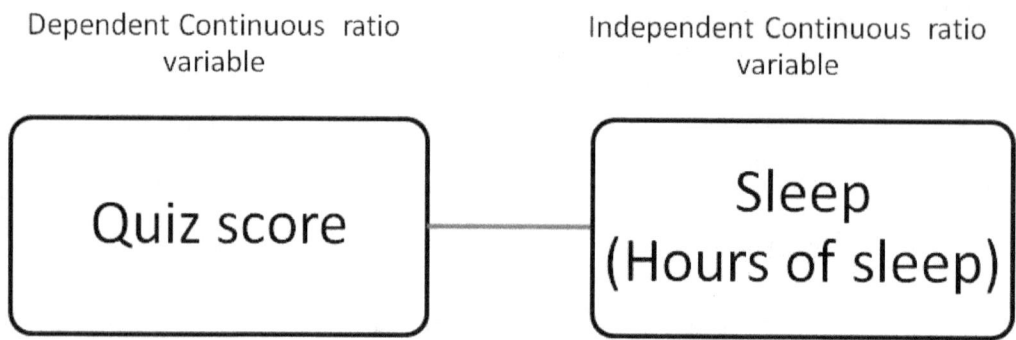

Figure 1.6: Example 1
In Figure 1.6, we want to see how sleep affects quiz score. Both variables are ratio which means there is an absolute zero and infinite number of possible values. Figure 1.7 is for example 2 and depicts a independent categorical variable and a dependent continuous variable.

- Question-Is there a relationship between the amount of sleep students receive and their performance on a quiz?

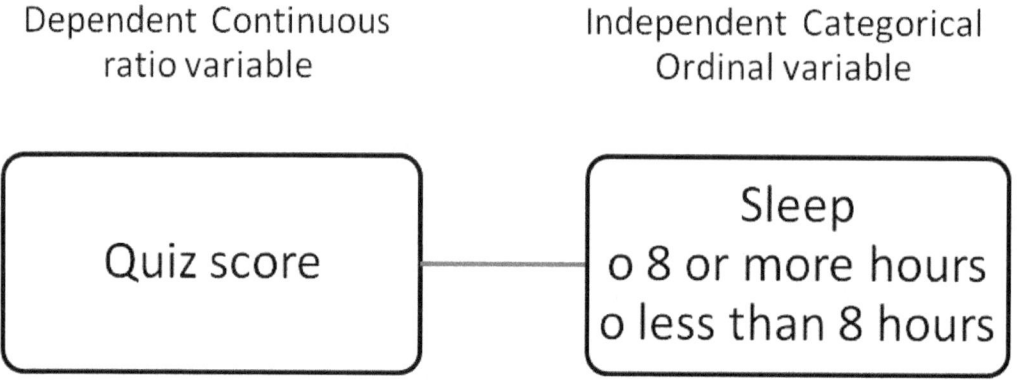

Figure 1.7: Example 2
Sleep this time has two possible options either more than 8 hours or less than 8 hours. There are no other possible values. Quiz score however is the same as in example 1. Example 3 shows an independent continuous variable influencing a dependent categorical variable.

- Question-Is there a relationship between the time students spend studying and whether the pass or fail a course.

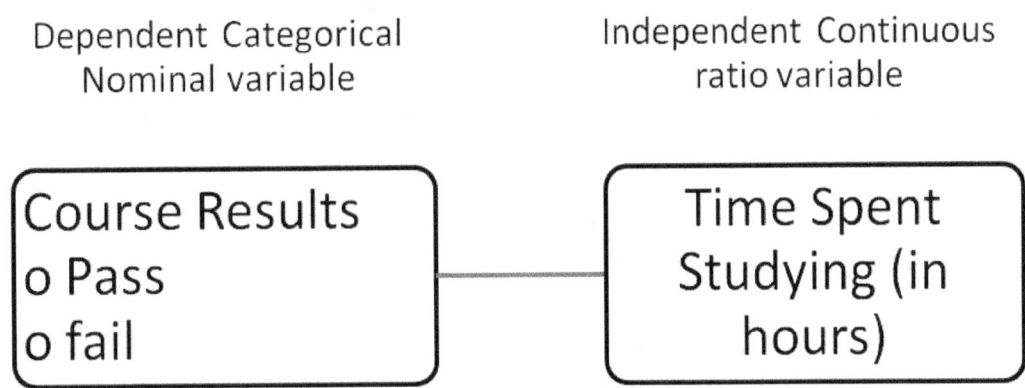

Figure 1.8: Example 3

This is example allows time spent studying to be any value while course results are limited to pass or fail. Figure 1.9 shows example 4 below and provides an example of a categorical independent variable influencing a categorical dependent variable.

- Question-Is there a relationship between the program a student studies in and whether they pass or fail a course?

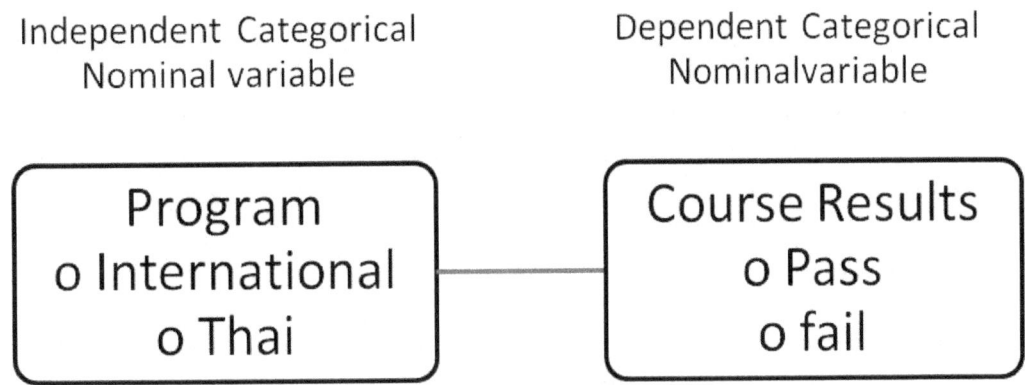

Figure 1.9: Example 4

The choices are limited on both sides as you can see in the examples. Essentially, these are the only possible relationships that we can look at in statistics. It does get more complicated with multiple independent variables but there is almost always one dependent variable that is either categorical or continuous in nature.

CHAPTER 1. WHAT IS STATISTICS?

Statistical Notation

Statistical notation is probably one of the characteristics of statistics that makes it so mysterious. In this section, we will try to unwrap the mystery behind it. Below is an example of statistical notation in Figure 1.10.

$$\sum_{i=1}^{n} X_1$$

Figure 1.10: Statistical Notation

Let's examine this piece by piece

- The Σ sign means "summation" it means add everything together when you are done.

- 'i' stands for the individual row in the data. So here, it means start with 1 and go up 1 for each row of data.

- n represents the size of the dataset. Therefore, we start at i or 1 and we don't stop adding until we get to n which is the last row in the data.

- X_1 indicates what exactly we are doing to each row of data. In this particularly example, we are simply adding each row of dating together. This will make more since in the example below We will now go through several examples together below.

Example 1.1

Below is some statistical notation. The expanded form of it is shown to the right of the equal sign. By expanded we mean what the statistical notation literally means

$$\sum_{i=1}^{2} X_i = X_1 + X_2$$

11

1. Start with the first row or example (i = 1)
2. Stop at the second example (2)
3. Put the values next to each other (X_i)
4. Add them all together (Σ)

We cannot finish summing the values because we do not have actual values for X yet in our example. In the example below we have a values for X that we can insert into our equation.

Example 1.2

Below are the values for X
$X_1 = 2$ $X_2 = 4$ $X_3 = 1$

Below is the equation with the answer

$$\sum_{i=1}^{3} X_i = X_1 + X_2 + X_3 = 2 + 4 + 1 = 7$$

All we did was plug in the three values of X. The letter *i* says to start it 1 and the number 3 above the summation means to stop at the third value of X. The value X_1 tells us just to add each value together. Let us do a slightly more complicated example with the same values for X in example 1.3.

Example 1.3

$$\sum_{i=1}^{3} X_i^2 = X_1^2 + X_2^2 + X_3^2 = 2^2 + 4^2 + 1^2 = 4 + 16 + 1 = 21$$

This is the same example. The only difference is the call to square the X values. Therefore, we first square the examples before adding them. This is important. The summation is always the last thing you do. All other commands are executed first. Example 1.4 is an example with more than one variable.

Example 1.4

Below are the values for X and also Y. Simplify the expression.

$X_1 = 2 \quad X_2 = 4 \quad X_3 = 1$
$Y_1 = 8 \quad Y_2 = 5 \quad Y_3 = 3$

$$\sum_{i=1}^{3}(X_i+Y_i)^2 = (X_1+Y_1)^2+(X_2+Y_2)^2+(X_3+Y_3)^2$$
$$=(2+8)^2+(4+5)^2+(1+3)^2$$
$$=10^2+9^2+4^2$$
$$=100+81+16$$
$$=197$$

This is hopefully becoming clearer. The only thing that is new is the introduction of a second variable. Notice how X_1 is grouped with Y_1 and the 2,s with the 2,s etc. Also notice that we first dealt with the information on the inside of the parentheses, we then squared the result and finally we added them together.

The purpose of statistical notation is to summarize what is happening mathematically. Instead of writing out all the individual values that need to be calculated and summed the notation describes this efficiently using Greek letters and variables.

Conclusion

Statistics is not difficult if you learn the characteristics of it systematically. Descriptive and inferential statistics play different roles in explaining numerical data. The level of measurement limits what you can know about your data while also allowing you to find answers to your questions. There are also times where variables are independent or dependent. Lastly, statistical notation is critical to understanding equations as we move forward.

Points to Remember

- There are several forms of variables

- There are two main types of statistics

- Variables can be independent or dependent
- Statistical notation explains what an expression is doing

Exercises

1.1

Direction: Determine if each example is a categorical or continuous variable

1. Number of motorbikes on campus
2. Most popular brands of motorbikes at AIU
3. GPA
4. Hours of sleep for each student
5. Average price of food in the cafÃl'
6. Hair color of faculty
7. Number of students on academic probation

1.2

Direction: Determine whether each example is measured at the nominal, ordinal, interval, or ratio level

1. Average salary of daily worker
2. Mother tongue
3. World Cup rankings
4. Kilometers per hour
5. Temperature in Celsius
6. Academic major
7. Nationality of students

CHAPTER 1. WHAT IS STATISTICS?

1.3

Direction: Write the following expressions in expended form

1. $\sum_{i=1}^{3}(2X_i+1)$

2. $\sum_{i=1}^{4}2(X_i+2Y_i)$

Used the values below to complete items 3 and 4

$X_1 = 1 \quad X_2 = 4 \quad X_3 = 3$
$Y_1 = 4 \quad Y_2 = 2 \quad Y_3 = 5$

3. $\sum_{i=1}^{3}(X_i^2+2Y_i)$

4. $\sum_{i=1}^{3}2(X_i+3Y_i)$

Chapter 2

How do You Use R?

In this chapter, you will learn the following...

- How to download and install R and RStudio.
- Basic coding concepts for using R.
- Introduction to functions.

Key Terms.

Descriptive statistics	Inferential statistics	Categorical variable	Continuous variable
Nominal variable	Ordinal variable	Interval variable	Ratio variable

Introduction

This chapter will explain how to install and use R. If you are not aware, R is a free statistical program that is available for download on the internet. It is one of the most-popular statistical software available.

This is not a book that is focused on the many details of R. Our main goal here is to learn enough about the language in order to do basic statistical operations. Only what is necessary to learn will be taught and explained.

In order to use R you must also install RStudio. RStudio is what we call an integrated development environment. This program allows you to use the R programming language in a convenient manner. Therefore, we have two things we need to install and they are R and RStudio. The directions below are for windows-based PCs.

Installing R

To download R you need to go to the following website.

$$\text{https://cran.r-project.org}$$

This is the home page for "The Comprehensive R Archive Network" which is a resource for R. On the homepage, you will see link for "Download R for Windows" as shown below

CHAPTER 2. HOW DO YOU USE R?

On the next screen, you want to click on "base" as shown in the screen below

Lastly, you want to click on "Download R" as shown in the screen below

The rest of this process is controlled by windows and is similar to any other program you have downloaded and installed.

Installing RStudio

To install RStudio go to the link below

```
https://www.rstudio.com/
```

When you follow this link, it will take you to the following page.

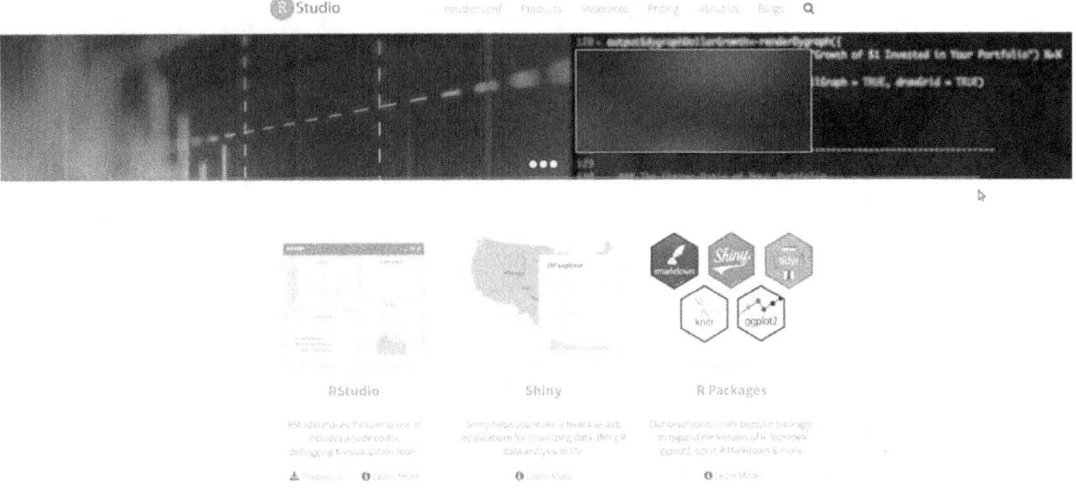

CHAPTER 2. HOW DO YOU USE R?

Next, you need to click on "download" under "RStudio" graphic. You will then see the following screen.

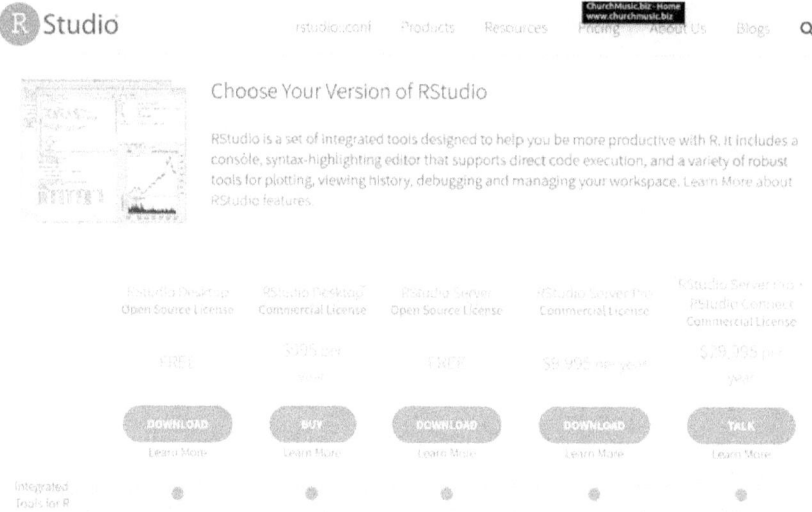

You need to download the "RStudio Desktop Open Source License" which is free. This choice is available to the far left. When you click on this button, you will be move to the bottom of the screen or a new tab will open. Whatever happens you want to install the RStudio version that looks similar to the following.

```
RStudio 1.1.383 - Windows Vista/7/8/10
```

The version might be different but you want the Windows version. When you click on this, the process will be similar to your experience downloading other programs from the internet.

R Basics

You must have R and RStudio installed in order to use the program successfully. When you want to use R, you need to open RStudio. For windows, just click on the start button to find RStudio. When you run RStudio, you should see the following although there may be variations in the color scheme.

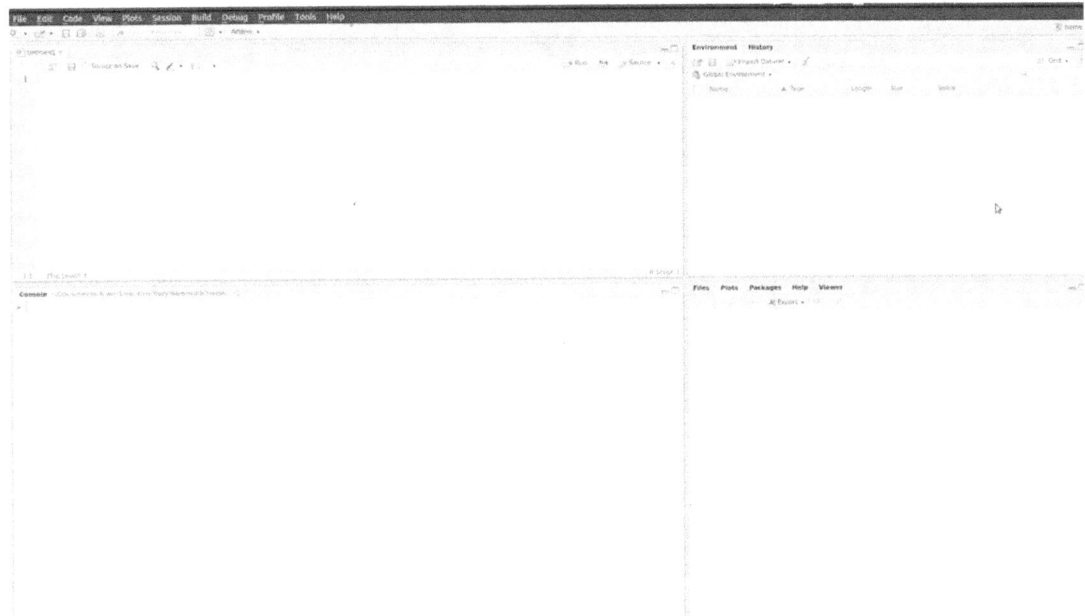

- The top left box is the Script Pane-This pane is used to run multiple lines of code. You can save your code here for future use as well.

- The bottom left box is the Console Pane-In this pane you can run a single line of code at a time. In addition, all outputs from R are printed here.

- The top right box is the Workspace-This is where different information is stored such as the objects in the environment. If this does not make sense do not worry about it.

- The bottom right box is the Viewer pane-This pane allows you to see the output of graphs as well as to search for help.

Using R

You can do all the basic calculations that a calculator can do in R. Below are examples using the console. The line with the > is the code or input. The line with the brackets symbol is the output. To run any code you need to press enter. Later, we will use the script pane. Below are some basic operations you could do in a calculator or in Excel.

Addition
> 2+2
[1] 4
Subtraction

```
> 2-2
[1] 0
```
Multiplication
```
> 2*2
[1] 4
```
Division
```
> 2/2
[1] 1
```
Notice the use of the * and / for multiplication and division.
Powers
```
> 2 ^2
[1] 4
```

You can also save information inside things called "objects." These are variables that can be used for additional calculations. Below is an example. First we will save the number 2 in the object called "wow." Then we will print the information to the console in R by typing the name of the object and pressing enter.
```
> wow<-2
> wow
[1] 2
```

Whenever you make an object you must put the name of the object followed by <- to save the information. As shown below

```
                Object name <- stuff you want to store
```

You can add objects together like numbers most of the time as well.
```
> wow + wow
[1] 4
```
In the example above, since we set wow = 2. If we add wow + wow we will get 4. The two most common objects are vectors and dataframes. A vector is one row of data such as our "wow" object. A dataframe is an object with multiple rows of data such as in a Microsoft Excel spreadsheet. We will learn more about this throughout the book.

Functions

Functions are pieces of code that do something for you automatically. At a minimum a function has the following.

```
                Function name(arguments)
```

The function is on the outside of the parentheses. Inside you have the object as well as any arguments. Below in Figure 2.1 is a visual of a function.

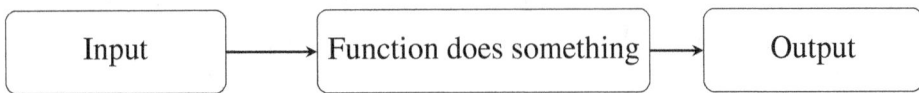

Figure 2.1: Function Diagram

We are going to learn three functions for practice below and they are the sum, prod, and c function.

sum function

The sum function adds together all the arguments inside the parentheses. For example,
> sum(2,3,4,5)
[1] 14

The answer is 14 or the same as 2 + 3 + 4 + 5 = 14. The sum function automatically adds the numbers that are separated by commas inside the parentheses. Figure 2.2 is a diagram of what the sum function does. Next we have the prod function.

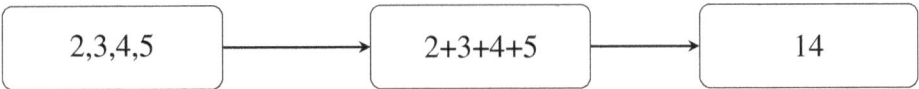

Figure 2.2: sum fuction

prod function
> prod(2,3)
[1] 6

Above we get the answer of 2 * 3 = 6 using the prod function. Figure 2.3 is a diagram of the prod function. The c function is the last function we will discuss for now.

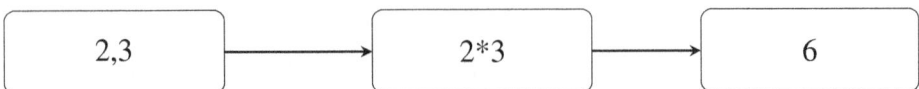

Figure 2.3: prod function

c function

The c function combines information into a single vector of information that is usually saved in an object. c stands for concatenate. Below is an example.
>what<-c(10,2)

CHAPTER 2. HOW DO YOU USE R?

```
> what
[1] 10 2
```

By using the c function we were able to save the numbers 10 and 2 in the object we created called "what." It is also important to remember that R is case-sensitive. In other words, love and Love are different in the R language. Therefore, you must be careful in how you type code. Figure 2.4

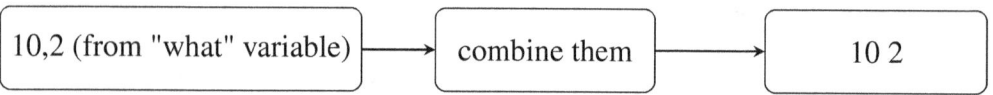

Figure 2.4: Function Diagram

Loading Default Data

There are many pre-installed datasets in R. The one we will use for the majority of the book is the iris dataset. To see the iris dataset type the following and press enter.

```
> iris
```

To see what variables are in the dataframe use the `str` function as shown below.

```
> str(iris)
'data.frame':    150 obs. of  5 variables:
 $ Sepal.Length: num  5.1 4.9 4.7 4.6 5 5.4 4.6 5 4.4 4.9 ...
 $ Sepal.Width : num  3.5 3 3.2 3.1 3.6 3.9 3.4 3.4 2.9 3.1 ...
 $ Petal.Length: num  1.4 1.4 1.3 1.5 1.4 1.7 1.4 1.5 1.4 1.5 ...
 $ Petal.Width : num  0.2 0.2 0.2 0.2 0.2 0.4 0.3 0.2 0.2 0.1 ...
 $ Species     : Factor w/ 3 levels "setosa","versicolor",..: 1 1 1 1 1 1 1 1 1 1 ...
```

As you can see, we have 150 rows of data and 5 variables. If you want to look at a specific variable use the dataframe name followed by the dollar sign to looker at it.

```
> iris$Sepal.Length
[1] 5.1 4.9 4.7 4.6 5.0
```

The variable above is the `Sepal.Length` variable from the iris dataset. If you look at the output it is the same as the $ Sepal.Length: num... information under the `str(iris)` code shown above.

Loading Your Own Data

It is probably much more common to load and use your own data. To load you own data that you saved as a CSV do the following...

1. Click on "import dataset" in the workspace panel

2. Select "from CSV"

3. Find your file on your computer

4. Click "import"

How to Think Like a Coder

In order to be successful using R you must learn how to think like a coder. When you first are learning to analyze with R it is important to write down in advance what it is you want to do with your code. You must write step-by-step what you want to do. If you simply want to just make things up along the why you will quickly get confused and lost. Coding requires planning. To encourages this, this book will include several diagrams that will provide a picture of what was just coded.

Conclusion

This chapter was a break from dealing with the details of statistics. However, it was critically important for you to become familiar with R and RStudio. Most statistical analysis is done using computers. The only reason to do things manually is to understand what the computer is doing. In this book, we will learn how to calculate manually most of the concepts before learning how to code it in RStudio.

Points to Remember

- R is a statistical programming language that is free

- RStudio is an integrated environment that runs R

- Objects and functions are basics components to use in R

R Coding Used

- sum(): gives you the sum of the numbers in the parentheses
- prod(): gives you the product of the numbers in the parentheses
- c():concatenates or combines the values in the parentheses into a vector
- str(): Provides a brief glimpse of an object
- The $ is used to access a single piece of data in an objects

Exercises

2.1

Directions: Calculate the following using R

```
1. 45 + 67

2. 34 + 677

3. 100 * 123

4. 7868 - 456

5. 450 / 10

6. 23^4

7. 25^8
```

2.2

Directions: Calculate the following using R

```
1. Cool <- 5

2. Cool + Cool

3. Cool * Cool

4. Cool -10
```

5. Cool * 100

6. Cool / 10

7. Cool^3

2.3

Directions: Use the functions below in R

1. sum(2,4,6,8)

2. sum(12,45,567,99)

3. prod(5,7,6)

4. prod(9,4,2)

5. cool<-c(10, 4)

6. prod(cool)

7. sum(cool)

8. sum(cool) / prod(cool)

2.4

Directions: Use the functions below in R

1. Use the str function on the cars dataset

2. Look at the speed variable only in the cars dataset using the $

Chapter 3

How do You Visualize Numbers

In this chapter, you will learn the following...
- The purpose of data visualization.
- How to make frequency tables
- How to make histograms, bar graphs, pie charts, scatter plots
- Export visuals to Word

Key Terms.
  Frequency tables   Class limits   Class interval
  Stem and leaf plot Histogram   Scatter plot

Introduction

Once data is collected it needs to be organized and placed into rows and columns. This is normally done using Microsoft Excel for small projects and a database for larger projects. The way to get raw data from surveys into a computer is beyond the scope of this book. However, once the data is in the computer some of the first things that you do are look at the frequency distribution and create several different visualizations, such as plots and graphs, in order to see what the data looks like. Seeing the data in visualizations helps you decide what to do with it.

Data visualization is powerful because what use to be a bunch of strange numbers is clearly explained by simply placing dots between two axis. Through doing this, it allows you to see trends, patterns, and relationships in the data, as well as the ability to draw and make inferences from the data. This is perhaps why in English we have the idiom "A picture is worth a thousand words."

In this chapter, we will look at several commonly used data visualization tools. For beginners, this is actually some of the more complex R code in this book. However, data visualization is a critical first step and is often one of the first concepts presented in a statistics textbook. Therefore, even though the coding is somewhat complex it is important to develop this skill first as a researcher.

Frequency Table with Categorical Variable

A frequency table tells you how frequent a particular value is in a dataset. Data needs to be grouped in order for it to be organized and placed into frequency tables. When the data is categorical, you simply place the data into the table by the category. When the data is continuous, the process is slightly more complicated and involves the creation of bins that have a range of numbers in them. We are going to learn how to make a frequency table manually first. Then we will learn how to make one using R.

Learning how to make a frequency table by hand may seem pointless when a computer can do it. However, doing it the old-fashion way helps you to understand what

CHAPTER 3. HOW DO YOU VISUALIZE NUMBERS

is happening and this will improve your ability to interpret and manipulate data in the future. Below is our example for categorical variables.

Example 3.0 Table for Categorical Variables

Below is the results of the data collection of 10 students by gender. Create a frequency table with the information below.

Girl Boy Girl Girl Boy
Boy Girl Girl Boy Boy

There are three steps to completing this problem

1. Create table

2. Count the number of examples for each category

3. Determine the percentage of each category

Step: 1 We need to create a table that has two rows. A row for girls and a row for boys. We also need three columns. The first column will be our categories, which are based on gender. The second column will be the frequency of each. The last column will be the percentage of each category. Below is an empty table.

Category	Frequency	Percent
Girl		
Boy		

Step: 2 We first need to count the number of girls and boys. This is done for you in the table below.

Category	Frequency	Percent
Girl	6	
Boy	4	

Step: 3 The final step is to determine the percentages. In order to do this, you take the frequency of category divided by the total number of examples multiplied by 100. The general equation for this is below followed by the updated table.

$$percent = (\frac{Frequency}{total\ number\ of\ observations}) * 100$$

Category	Frequency	Percent
Girl	6	60%
Boy	4	40%

We are now going to make a frequency table using the iris dataset in R. We are specifically going to use the Species variable and count the number of each type of flower in the dataset.

The table above is what the final version looks like for our purposes. We will now turn our attention to learning how to create this same table in R.

Example 3.1: Using R to Create a Frequency Table

First, I would like you to see what the final product looks like.

```
##         Var1 Freq   percent
## 1     setosa   50 0.3333333
## 2 versicolor   50 0.3333333
## 3  virginica   50 0.3333333
```

As you look at the table above you can see that the frequency table has three columns of data. The first column is the variable that tells you the name of the different species of flowers. The second column tells you how many of each flower there are in the data set. Lastly, the last column tell percentage of the total of each flower is.

We are now ready to develop the table. We are going to call our object Example3.1 because we are in chapter 3 and this is the first example using R. To make our table we will use the `table` function. Below is the code with the output followed by Figure 3.1 which is a diagram of what the function did.

```
Example3.1<-table(iris$Species)
Example3.1
##
##     setosa versicolor  virginica
##         50         50         50
```

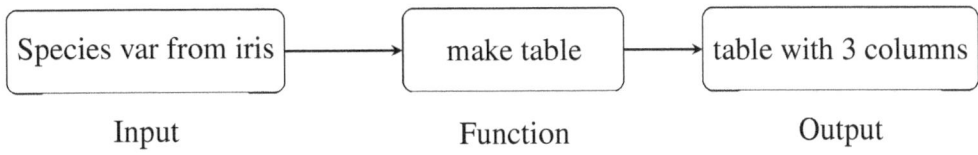

Input   Function   Output

Figure 3.1: table function

As you can see, we have the correct information but it does not look like our example. The data is in one row rather than three. Therefore, we need to find a way to transpose the data so that it is in three rows rather than one.

One way to do this is to use the `data.frame` function on our Example 3.1 as shown below followed by a diagram of what the function did.

```
Example3.1<-data.frame(Example3.1)
Example3.1
##          Var1 Freq
## 1      setosa   50
## 2  versicolor   50
## 3   virginica   50
```

```
┌─────────────┐    ┌──────────────────────┐    ┌──────────────────────────┐
│ Example 3.1 │ →  │ Change to data frame │ →  │ data frame with 2 columns│
└─────────────┘    └──────────────────────┘    └──────────────────────────┘
     Input                 Function                      Output
```

Figure 3.2: table function

Beautiful! We already have two columns of the table. Now we need to create our percent column.

To make the percent column, we need to make a separate object called "percent". This object will use the `table` function using the species variable from the iris dataset in order to find the proportions of each species. You have already seen this so we will not look at it again.

After we store the species in the percent table, we now need to calculate the proportions of each species using the `prop.table` function. When this is done, we will call the percent object so you can see what it looks like.

```
percent<-table(iris$Species)
percent<-prop.table(percent)
percent
## 
##     setosa versicolor  virginica 
##  0.3333333  0.3333333  0.3333333
```

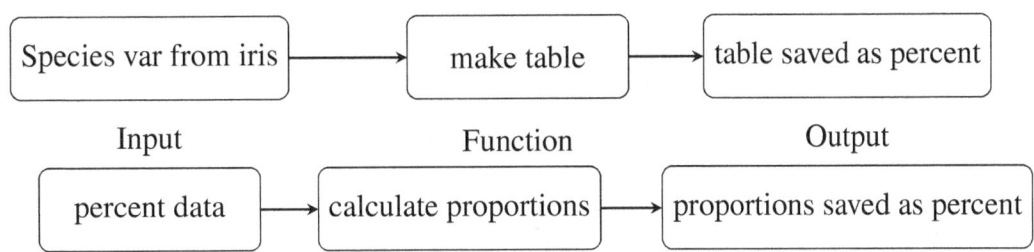

Figure 3.3: table & prop.table function

You can see we have our information. Now we will add this information to our Example3.1 object. This will be done by using the $ sign. When you type the object and put a $ sign after it it will allow you to add a column to your data. If this is hard to understand just follow the code below to see the results of it.

```
Example3.1$percent<-percent
Example3.1
##          Var1 Freq   percent
## 1      setosa   50 0.3333333
## 2 versicolor   50 0.3333333
## 3  virginica   50 0.3333333
```

By using the dollar sign we were able to add the column percent to our Example 3.1 and complete our table. This concludes how to create a frequency table categorical data. We will now look at how to make a frequency table with continuous data.

Frequency Table with Continuous variable

The unique problem with frequency tables for continuous variables is that you have to determine the following.

- The number of categories
- The interval of each category

By interval, we mean something such as 2-5 or 6-9. In other words, the numerical distance or range of each category. Just as a note, sometimes the word "class" is used instead of "category" and we will use both terms interchangeable.
To calculate the number of categories we will use the "2k rule". To determine the number of classes you should take a value where

$$2^k > \text{total number of observations}$$

This information can then be used to determine the class intervals. If this is confusing we will do it by hand first in Example 3.2 below.

Example 3.2

Below are the quiz results for a statistics class. Create a frequency table with the information provided.

$$\begin{array}{cccccccccc}
74 & 83 & 84 & 73 & 80 & 68 & 64 & 87 & 74 & 87 \\
87 & 73 & 68 & 93 & 82 & 69 & 90 & 57 & 88 & 83 \\
61 & 75 & 77 & 84 & 78 & 60 & 65 & 80 & 76 & 83
\end{array}$$

The following steps will be taken

1. Determine the number of classes/categories
2. Determine the class intervals
3. Create the class limits
4. Create the class boundaries
5. Calculate frequencies
6. Calculate the percentages
7. Determine cumulative percentages
8. Calculate the midpoints

Step 1: Determine the number of classes.
Using the 2^k rule we get the following. Focus on the answer rather than the math. Remember that this is done for you automatically in R. The number 30 below is the size of the sample.

$$2^k > 30$$

$$log(2^k) > log(30)$$

$$k * log(2) > log(30)$$

$$k > \frac{log(30)}{log(2)}$$

$$k > 4.9$$

Therefore, we need 5 classes.

Step 2: Determine class interval

We need five classes or categories. However, we do not know the range that each class should cover. Therefore, we need to calculate the class interval. The class interval involves taking the highest value subtracted from the lowest values divided by the number of classes below is the equation.

$$Class\ interval = \frac{highest\ value - lowest\ value}{number\ of\ classes}$$

Our sample is small at only 30 observations so we can find the highest and lowest values just by looking at the table. It appears that the highest value is 93 and the lowest is 57. Remember that we have already calculated the number of classes at 5. Therefore, our equation is completed below.

$$Class\ interval = \frac{highest\ value - lowest\ value}{number\ of\ classes} = \frac{93 - 57}{5} = \frac{36}{5} = 7.2$$

We should have intervals of about 8 as it is usually better to round up. What this means is that the five categories we are going to use should each have a range of 8.

Step 3: Class limits

To create the class limits, which is the actual range of each class, you start from the lowest value and count by the class interval until you reach the highest value. This has been done for you in the table below.

Class Limits
57 - 64
65 - 72
73 - 80
81 - 88
89 - 96

We start with the value 57 because this is the lowest value in our dataset. We then count up 8 to 64. We repeat this process 5 times because we originally determined we needed 5 classes.

Step 4: Class Boundaries

The class boundaries are always .5 below the lower level of a class interval and .5 above the highest value of a class interval this is useful when you have data with decimals.

CHAPTER 3. HOW DO YOU VISUALIZE NUMBERS

Class Limits	Class Boundaries
57 - 64	56.5 - 64.5
65 - 72	64.5 - 72.5
73 - 80	72.5 - 80.5
81 - 88	80.5 - 88.5
89 - 96	88.5 - 96.5

Step 5: Count the Frequencies
Now we count how many observations fall within each class.

Class Limits	Class Boundaries	Frequency
57 - 64	56.5 - 64.5	4
65 - 72	64.5 - 72.5	4
73 - 80	72.5 - 80.5	10
81 - 88	80.5 - 88.5	10
89 - 96	88.5 - 96.5	2

Step 6: Percentages
We calculate the percentages as shown below.

Class Limits	Class Boundaries	Frequency	Percent
57 - 64	56.5 - 64.5	4	(4 / 30) * 100 = 13
65 - 72	64.5 - 72.5	4	(4 / 30) * 100 = 13
73 - 80	72.5 - 80.5	10	(10 / 30) * 100 = 33
81 - 88	80.5 - 88.5	10	(10 / 30) * 100 = 33
89 - 96	88.5 - 96.5	2	(2 / 30) * 100 = 6.7

Step 7: Cumulative Percent
The cumulative percent is the running total of the percent it is shown below.

Class Limits	Class Boundaries	Frequency	Percent	Cumulative Percent
57 - 64	56.5 - 64.5	4	(4 / 30) * 100 = 13	13
65 - 72	64.5 - 72.5	4	(4 / 30) * 100 = 13	13 + 13 = 26
73 - 80	72.5 - 80.5	10	(10 / 30) * 100 = 33	13 + 13 + 33 = 59
81 - 88	80.5 - 88.5	10	(10 / 30) * 100 = 33	13 + 13 + 33 + 33 = 92
89 - 96	88.5 - 96.5	2	(2 / 30) * 100 = 6.7	13 + 13 + 33 + 33 + 6.7 = 98.7

There is some rounding error but the total is mostly there.

Step 8: Calculate the Midpoints
We can also find the midpoints. The midpoints is the middle value of each class limit. It is calculated below.

Class Limits	Class Boundaries	Frequency	Percent	Cumulative Percent	Midpoint
57 - 64	56.5 - 64.5	4	(4 / 30) * 100 = 13	13	(57 + 64) / 2 = 60.5
65 - 72	64.5 - 72.5	4	(4 / 30) * 100 = 13	13 + 13 = 26	(65 + 72) / 2 = 68.5
73 - 80	72.5 - 80.5	10	(10 / 30) * 100 = 33	13 + 13 + 33 = 59	(73 + 80) / 2 = 76.5
81 - 88	80.5 - 88.5	10	(10 / 30) * 100 = 33	13 + 13 + 33 + 33 = 92	(81 + 88) / 2 = 84.5
89 - 96	88.5 - 96.5	2	(2 / 30) * 100 = 6.7	13 + 13 + 33 + 33 + 6.7 = 98.7	(89 + 96) / 2 = 92.5

You can clearly see that this is a lot of work to do manually. This is perhaps one reason why nobody does it this way anymore. However, you now know how this is done and the reasoning behind it. We will now proceed to do this using Rstudio. We will use the iris dataset and this time we will use the Sepal.Length variable for our analysis.

Example 3.3: Using R for Frequency Table with Continuous Variable

Step 1: Determine the number of class

Using the str function on the iris data set and you will see that there are 150 observations. If we use the 2^k rule it means we need 8 classes ($2^8 > 150$).

Step 2: Determine class interval & Step 3: Class Limit

To determine the class limits we need to find the highest and lowest value in our iris Sepal.Length variable. This is found in the code below using the range function.

```
range(iris$Sepal.Length)
## [1] 4.3 7.9
```

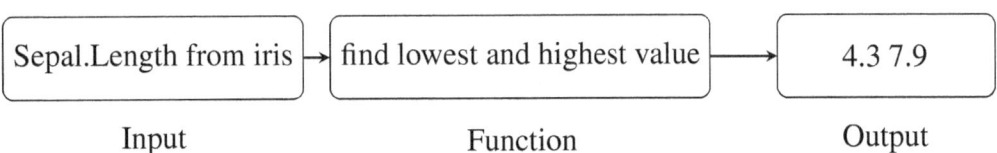

Input   Function   Output

Figure 3.4: range function

We take this information and find the difference between the two values. Next, we divide the answer by the number of classes we want. All of this is done below.

```
(7.9-4.3)/8
## [1] 0.45
```

Our class interval should be 0.5. This is because we will round the value. Rounding makes it easier to setup the needed classes. In addition, rounding aids interpretation for you and the audience.

Step 4: Class Boundaries & Step 5: Count the Frequencies

We need to divide the Sepal.Length variable. We do this with two functions the cut function and the seq function. The sequence function uses a start number, an end number as well as a number that tells it how many numbers to move by. To make this simple, the first number in our seq function will be 4 or about the lowest value in our variable. The highest number will be 8 or the highest number in our variable. Finally, the by number will be .5 which is the size of our class interval. The actual cut function simply makes this a categorical variable. This is hard to understand until you see the code below in Figure3.5. Figure 3.6 is a detailed visual of what the cut function does which is simply to divide a given range by a certain amount. We save all this in the object Example3.3

```
Example3.3<-cut(iris$Sepal.Length,seq(4,8,by=.5))
```

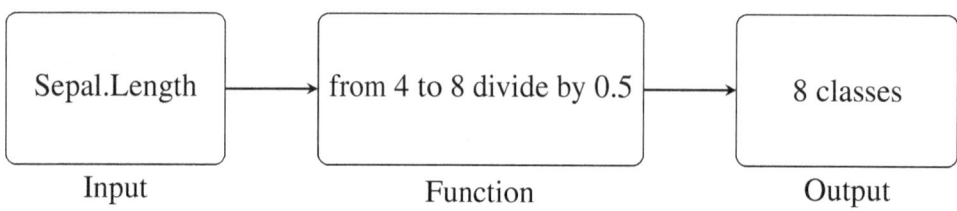

Figure 3.5: cut function

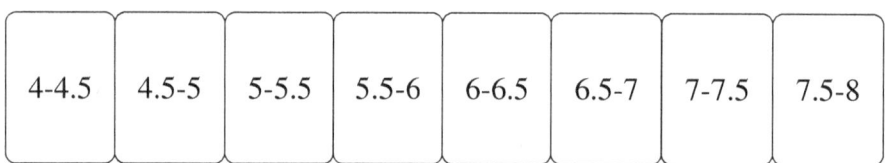

Figure 3.6: Details of cut function

The next two parts we have done before. First, we now need to convert Example 3.3 to a table. We will use the table function for this and we will call the object. This will allow us to see the classes and the frequencies of each class.

```
Example3.3<-table(Example3.3)
Example3.3
## Example3.3
##  (4,4.5] (4.5,5]  (5,5.5] (5.5,6]  (6,6.5] (6.5,7]  (7,7.5] (7.5,8]
##        5      27       27      30       31      18        6       6
```

We need this to have rows and not columns. Therefore our second move will be to use the data.frame function.

```
Example3.3<-data.frame(Example3.3)
Example3.3
##     Example3.3 Freq
## 1    (4,4.5]    5
## 2    (4.5,5]   27
## 3    (5,5.5]   27
## 4    (5.5,6]   30
## 5    (6,6.5]   31
## 6    (6.5,7]   18
## 7    (7,7.5]    6
## 8    (7.5,8]    6
```

We have a small problem. The first column in our data is called Example3.3 This is wrong and we need to rename it. We need to use the rename function from the dplyr package. A package is simply a collection of functions. You can download it using the code install.package("dplyr"). Once this is done you need to use the rename function, put the name of the object in the parentheses, give the new name for the column, place an equal sign and give the old name for the column. Follow the code below.

```
library(dplyr)
Example3.3<-rename(Example3.3,classlimit=Example3.3)
Example3.3
##     classlimit Freq
## 1    (4,4.5]    5
## 2    (4.5,5]   27
## 3    (5,5.5]   27
## 4    (5.5,6]   30
## 5    (6,6.5]   31
## 6    (6.5,7]   18
## 7    (7,7.5]    6
## 8    (7.5,8]    6
```

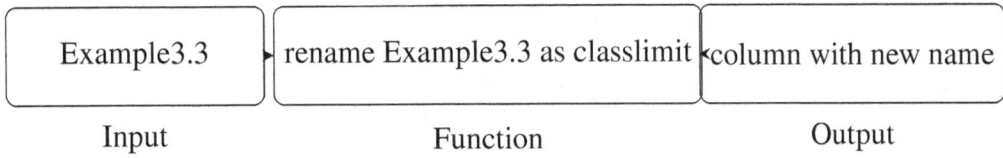

| Example3.3 | rename Example3.3 as classlimit | column with new name |
| Input | Function | Output |

Figure 3.7: rename function

Step 6: Percentage & Step 7 Cumulative Percentage

Next, we calculate the cumulative frequency. This is done using the `cumsum` function and the data comes from the freq column in our Example3.3 object. We will save this in our Example3.3 object in a new column called cumlativeFreq.

```
Example3.3$cumlativeFreq<-cumsum(Example3.3$Freq)
Example3.3
##   classlimit Freq cumlativeFreq
## 1    (4,4.5]    5             5
## 2    (4.5,5]   27            32
## 3    (5,5.5]   27            59
## 4    (5.5,6]   30            89
## 5    (6,6.5]   31           120
## 6    (6.5,7]   18           138
## 7    (7,7.5]    6           144
## 8    (7.5,8]    6           150
```

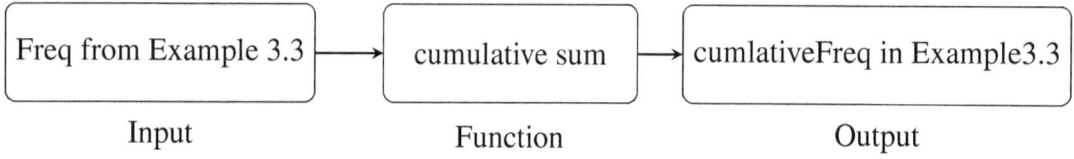

Figure 3.8: cumsum function

Now we find the percentages. We repeat the previous process but use the `prop.table` and create a new column called percent. We have done this before

```
Example3.3$percent<-prop.table(Example3.3$Freq)
Example3.3
##   classlimit Freq cumlativeFreq    percent
## 1    (4,4.5]    5             5 0.03333333
## 2    (4.5,5]   27            32 0.18000000
## 3    (5,5.5]   27            59 0.18000000
## 4    (5.5,6]   30            89 0.20000000
## 5    (6,6.5]   31           120 0.20666667
## 6    (6.5,7]   18           138 0.12000000
## 7    (7,7.5]    6           144 0.04000000
## 8    (7.5,8]    6           150 0.04000000
```

The cumulative percent is calculated with the code below.

```
Example3.3$cumulativePer<-cumsum(Example3.3$percent)
Example3.3
##   classlimit Freq cumlativeFreq   percent  cumulativePer
## 1    (4,4.5]    5             5 0.03333333    0.03333333
## 2    (4.5,5]   27            32 0.18000000    0.21333333
## 3    (5,5.5]   27            59 0.18000000    0.39333333
## 4    (5.5,6]   30            89 0.20000000    0.59333333
## 5    (6,6.5]   31           120 0.20666667    0.80000000
## 6    (6.5,7]   18           138 0.12000000    0.92000000
## 7    (7,7.5]    6           144 0.04000000    0.96000000
## 8    (7.5,8]    6           150 0.04000000    1.00000000
```

Step 8: Calculate the Midpoints

Lastly, we calculate the midpoints. It's difficult to explain this but basically we use the hist function to calculate the midpoint and then take this information and save it in our object. The code is below.

```
Example3.3$mid<-hist(iris$Sepal.Length)$mids
Example3.3
##   classlimit Freq cumlativeFreq   percent  cumulativePer  mid
## 1    (4,4.5]    5             5 0.03333333    0.03333333 4.25
## 2    (4.5,5]   27            32 0.18000000    0.21333333 4.75
## 3    (5,5.5]   27            59 0.18000000    0.39333333 5.25
## 4    (5.5,6]   30            89 0.20000000    0.59333333 5.75
## 5    (6,6.5]   31           120 0.20666667    0.80000000 6.25
## 6    (6.5,7]   18           138 0.12000000    0.92000000 6.75
## 7    (7,7.5]    6           144 0.04000000    0.96000000 7.25
## 8    (7.5,8]    6           150 0.04000000    1.00000000 7.75
```

Great work! This is how you make frequency table in R for continuous data. The remaining examples in this chapter are all done with computer code as they are never done manually ever these days. As such, things should move much faster and be easier from here on. We will now turn our attention to other forms of data visualization.

Stem and Leaf Plot

The steam and leaf plot is one of many ways to visualize data. It allows you to see the actual observations spread out. This type of plot is used for continuous data. Below is the code and visual is in Figure 3.9.

CHAPTER 3. HOW DO YOU VISUALIZE NUMBERS

```
stem(iris$Sepal.Length)
```

```
stem(iris$Sepal.Length)
##
##   The decimal point is 1 digit(s) to the left of the |
##
##   42 | 0
##   44 | 0000
##   46 | 000000
##   48 | 00000000000
##   50 | 0000000000000000000
##   52 | 00000
##   54 | 000000000000
##   56 | 00000000000000
##   58 | 0000000000
##   60 | 000000000000
##   62 | 0000000000000
##   64 | 000000000000
##   66 | 0000000000
##   68 | 0000000
##   70 | 00
##   72 | 0000
##   74 | 0
##   76 | 00000
##   78 | 0
```

Figure 3.9: Stem and Leaf Plot

What the stem and left plot tells you is how many of each value you have in your dataset. For example, we have one value of 4.2 in our dataset. It is represented as a 0 in the plot.

Histogram

Histograms are another way to see how data is spread out. They are used for continuous data. Figure 3.10 is a histogram.

```
hist(iris$Sepal.Length)
```

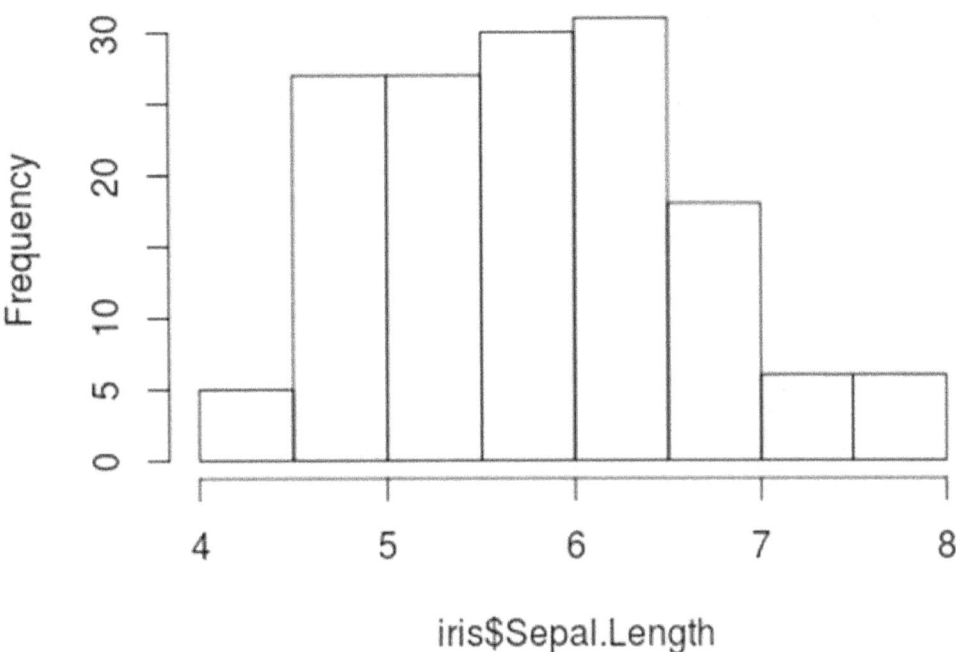

Figure 3.10: Histogram

Bar Graph

I am sure you are familiar with bar graphs. For this example, we use the species variable because bar plots are for categorical data. Notice how there is a small space between the bars. This is an indication that each bar is discrete or its own category.

```
bp<-table(iris$Species)
barplot(bp)
```

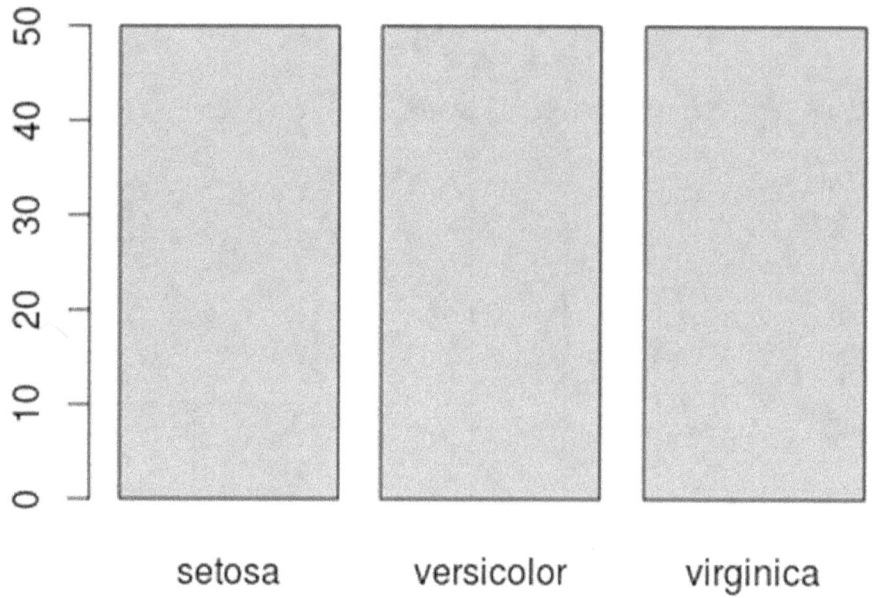

Figure 3.11: Bar Graph

Pie Graph

Pie graphs are another option. They to are used for categorical data. Pie graphs provide a visual of proportions. Figure 3.12 is the pie graph.

```
irispie<-table(iris$Species)
pie(irispie))
```

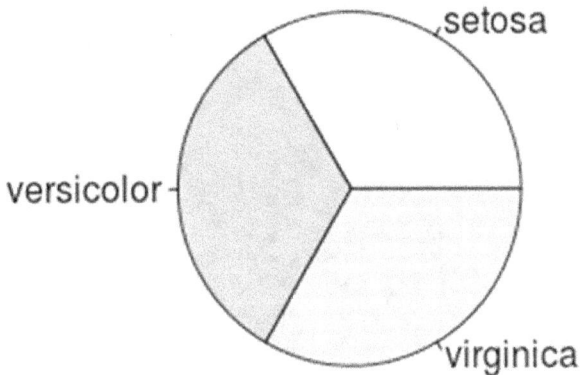

Figure 3.12: Pie Graph

Scatter Plot

A scatter plot is useful for looking at the relationship between two continuous variables. This becomes much more important when we discuss correlation and regression. Below is a scatter plot of Sepal.Length and Sepal.Width from the iris dataset.Figure 3.13 is the scatter plot.

```
plot(iris$Sepal.Length,iris$Sepal.Width)
```

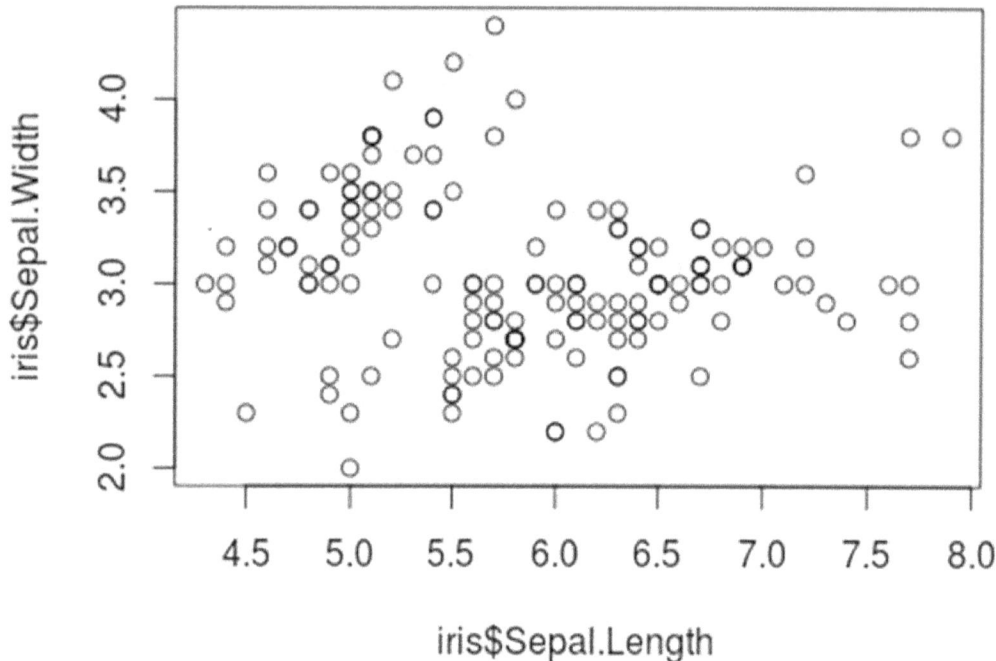

Figure 3.13: Scatter Plot

Exporting Figures to Word

For the table that we made at the beginning of the chapter, it is easy to move this to word for a paper. It is simply a matter of copying and pasting. However, the tables taken directly from R are not visually appealing. It is better to reformat the information by creating a table in Word and adding the information. For the typical paper, this is not a lot of work. For more extensive work other options are available that are not discussed here.

For the histograms, bar plots, scatter plots, etc. we will look at how to take these from R Studio to Word.

Step 1: Make the visual

Make the visual that you want to add to word. If you look at the screen you can see the histogram to the right.

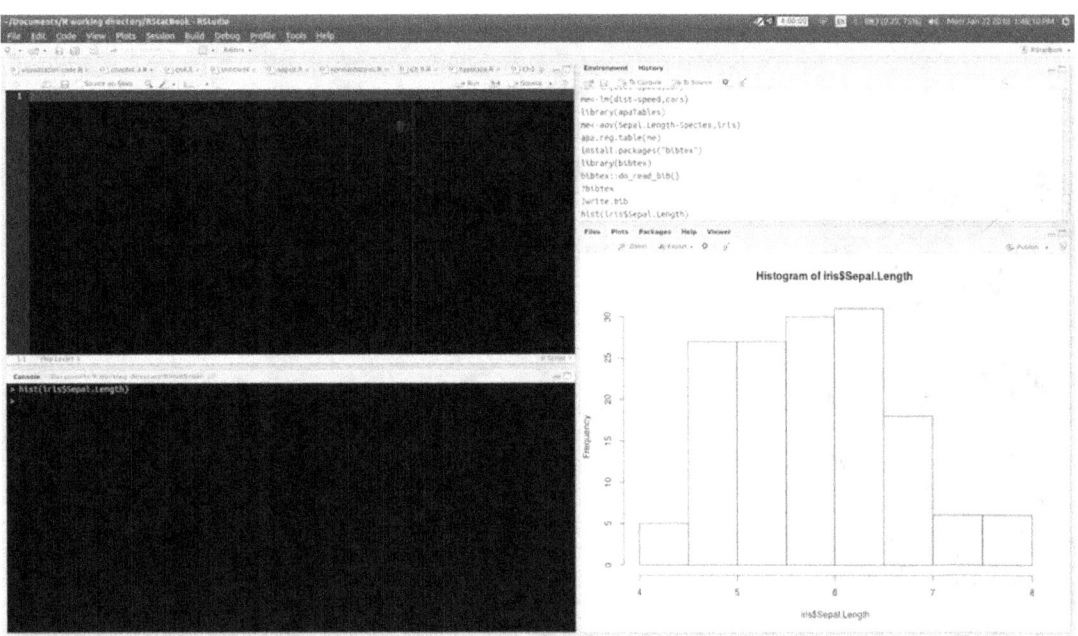

Figure 3.14: Step 1
Step 2: Export
To export the visual you need to click on export–>copy to clipboard

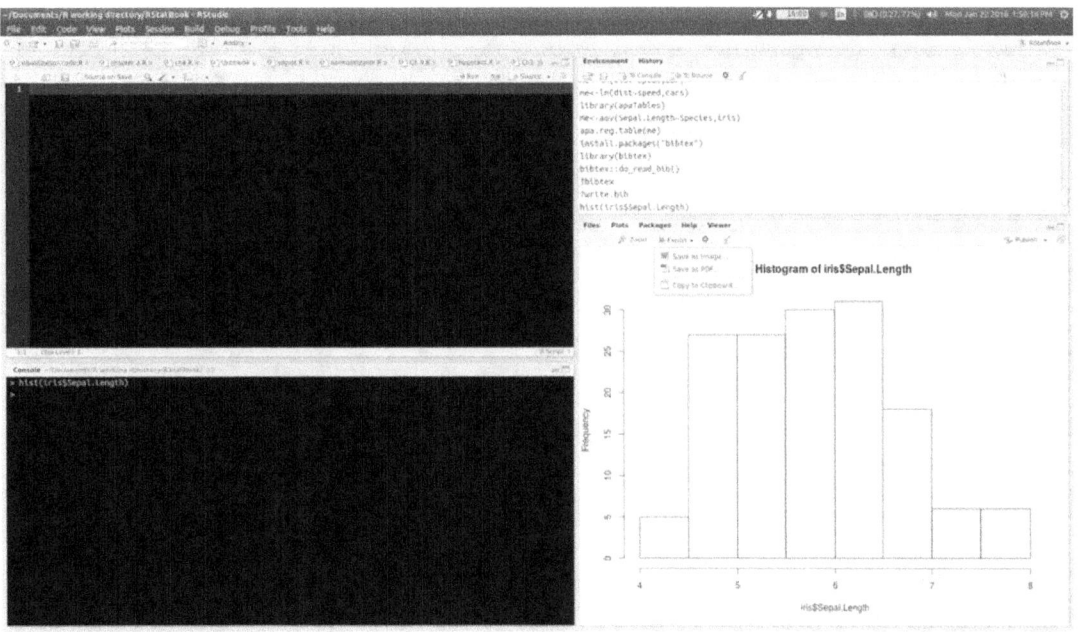

Figure 3.15: Step 2
Step 3: Determine Size
In the next screen, you can manipulate the size of the visual. When you are satisfied

CHAPTER 3. HOW DO YOU VISUALIZE NUMBERS

click "copy plot"

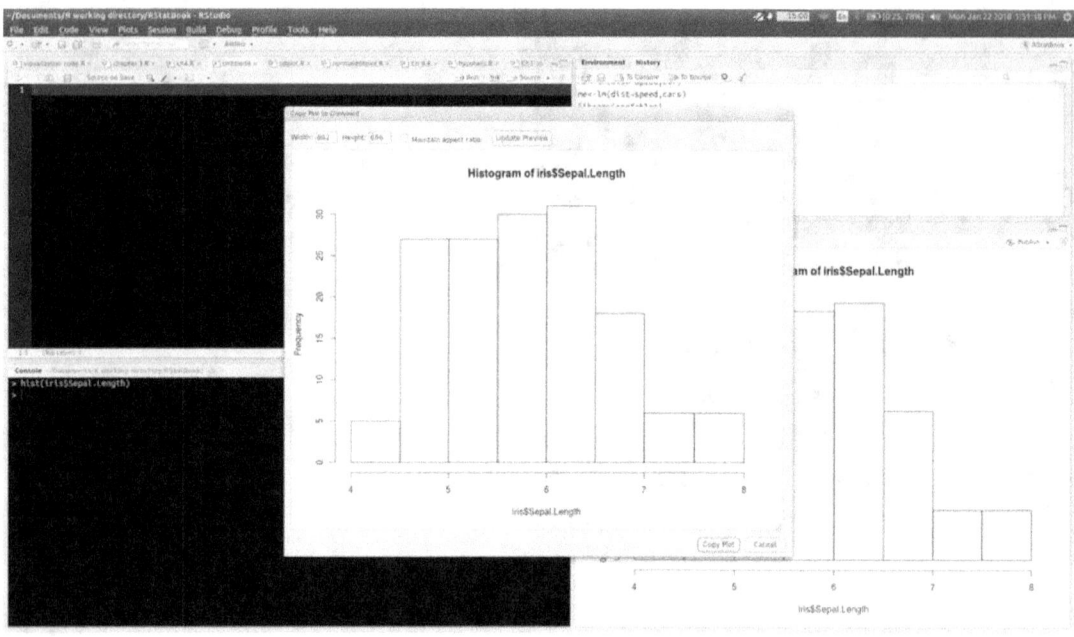

Figure 3.16: Step 3

Step 4: Paste
You can now paste the visual into word

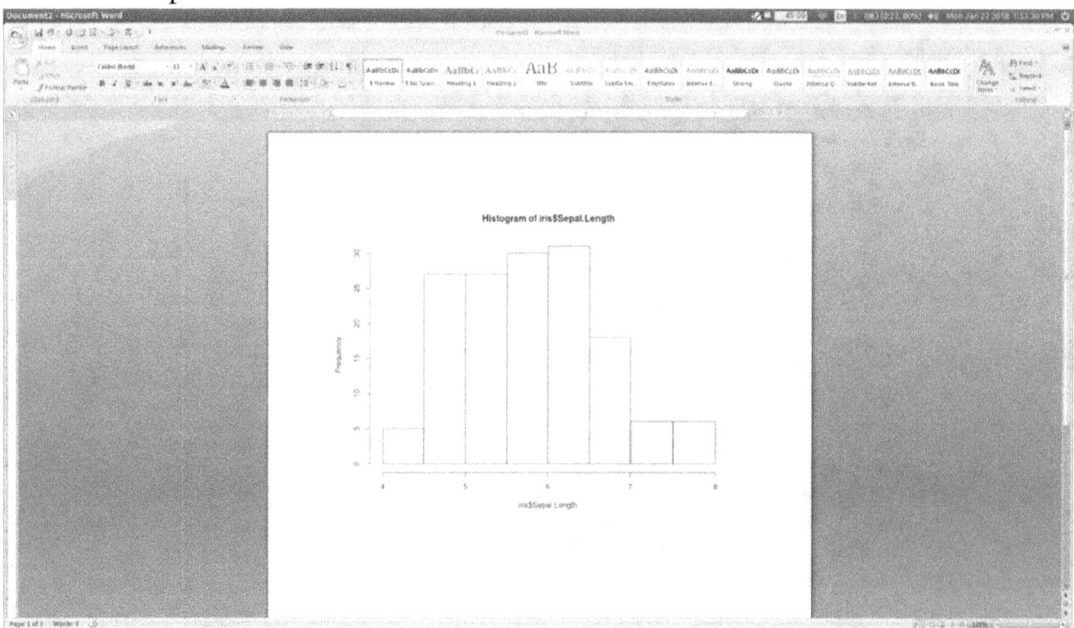

Figure 3.17: Step 4

Conclusion

Data visualization is a critical first step in statistical analysis. You need to see the data in order to determine if the analysis you planned to do is acceptable. There many different forms of data visualization you can use. The examples here are some of the most commonly used.

Points to Remember

- Data visualization is an important first step in analysis
- Tables are one form of summarizing data
- Other visualization tools include histograms and scatter plots

R Code Used

- table(): Makes a table of the data
- data.frame(): Converts an object to a data frame.
- range(): Finds the lowest and highest value in an object
- cut(): Divides a dataset into equal sizes you want.
- install.package(): Installs the package that is in quotes in the parentheses
- library(): Loads the package that is in parentheses
- cumsum(): Calculates the cumulative sum of an object
- hist(): Makes a histogram
- stem(): Makes a stem and leaf plot
- barplot(): Makes a bar graph
- pie(): Makes a pie graph
- plot(): Makes a scatter plot

Exercises

Directions: Create a frequency table from the following information

1.
good	bad	bad	bad	bad	good	good	good	good	bad
bad	bad	bad	bad	good	good	good	bad	bad	good

   You will need a table like the one below

Class	Frequency	Percent

2. Using R code, make a frequency table use the treatment variable from the OrchardSprays data set.

3. Make a frequency table for the continuous data below.

   48 56 43 49 52 46 59 49 44 45
   60 44 41 48 51 43 42 50 55 51

   You will need a table like the one below

Class	Class Boundary	Frequency	Percent	Cumulative Percent	Midpoint

4. Use the speed variable from the cars dataset to create a frequency table.

5. Make a stem and leaf plot of the `Sepal.Width` variable in the iris dataset

6. Make a histogram of the `Petal.Width` variable in the iris dataset

7. Make a pie graph of the treatment variable of the OrchardSprays dataset.

8. Make a bar graph of the variable diet in the ChickWeight dataset.

9. Make a scatter plot of the `Petal.Width` and `Petal.Length` variables in the iris dataset.

Chapter 4

What are Measures of Central Tendency

In this chapter, you will learn the following...

- Calculate several values of central tendency

Key Terms.
 Mean   Median   Mode

Introduction

Measures of central tendency are used to produce a single value that is helpful in finding the center of a dataset. One of the most commonly used measures of central tendency is mean. However, other examples include the median and the mode. The primary benefit of measures of central tendency is that they allow you to summarize many observations into a single number that is often an accurate representation of the dataset. Keep in mind that the mean can only be calculated for interval and ratio levels of measurement.

Mean

There are several different types of means such as arithmetic, geometric, weighted, etc. For our purposes, we will focus on arithmetic mean. Arithmetic mean uses all of the available values in a dataset equally to determine the mean. This the most commonly used form of mean in statistics.

To further complicate things, there are two levels at which the arithmetic mean can be calculated and they are at the population and sample level. The Population mean is used for the population and there is no need to develop inferences. The sample mean is used for samples and can be used to infer what the population mean is. The statistical expression is the same for both.

For a sample, the equation for mean is as follows.

$$\bar{x} = \frac{\sum x}{n}$$

$\bar{x}$ = *sample mean*
x = individual observation from the data set
$\sum$ = *sum all x values*
n = sample size

For the population mean, the equation is mostly the same.

$$\mu = \frac{\sum x}{N}$$

CHAPTER 4. WHAT ARE MEASURES OF CENTRAL TENDENCY

μ = sample mean
x = individual observation from the data set
Σ = sum all x values
N = Population size
Below is an example

Example 4.1

Below is a sample of the number of hours 10 students slept last night. Calculate the ungrouped arithmetic mean of the sample.

10 4 2 6 8 7 4 5 6 7

$$\bar{x} = \frac{10+4+2+6+8+7+4+5+6+7}{10}$$

$$\bar{x} = \frac{59}{10}$$

$$\bar{x} = 5.9$$

Find the Mean Using R

Below is how we would calculate the mean of the `Sepal.Length` variable in the iris dataset in R.

```
> mean(iris$Sepal.Length)
[1] 5.843333
```

Median

The median is the exact midpoint of a dataset. Primary benefit of median is that it is not influenced by extreme values as the mean is. In addition, median can be used for ordinal, interval, and ratio levels of measurement. Calculating the median is rather simple, if the sample size is an odd number the median is the number exactly in the middle as shown below

1, 2, 3 the median is 2

If the sample size is even you find the mean of the two middle values as shown below,

1, 2, 3, 4
Median = (2+3) / 2 = 2.5

Find Median Using R

Below is how you find the median in R using Sepal.Length variable from the iris dataset.

```
> median(iris$Sepal.Length)
[1] 5.8
```

This number 5.8 means that half the values are above 5.8 and half the values are below 5.8. The number 5.8 is the middle value or the median.

Mode

The mode is the value(s) that appear most frequently in a dataset. It is possible for there to be more than one mode. If two values are most common then it is bimodal, if three or more it is multimodal, and if only one, it is unimodal. Below is an example of unimodal.

Below is a sample of the number of hours 10 students slept last night. Calculate the find the mode of the sample.

$$10 \quad 4 \quad 2 \quad 6 \quad 8 \quad 7 \quad 4 \quad 5 \quad 6 \quad 7$$

Of all the numbers above only the number 7 appears more than once. As such, 7 is the mode Doing this in R requires somewhat complicated coding so we will not go through a demonstration of this.

Conclusion

The mean, median, and mode are foundational statistical tools for understanding your data. The mean in particular is used in a huge number of more advance statistical applications. Therefore, it is critical that you understand how measures of central tendency work as we progress further in our journey.

Points to Remember

- Mean provides a general sense of the midpoint of a dataset but is influenced by extreme values

- Median gives you the exact midpoint regardless of extreme values

- Mode is a measure of the most frequently occurring value in a dataset.

R Code Used

- median(): Calculates the median of the object
- mean(): Calculates the mean of an object

Exercises

4.1

1. Find the mean, median, and mode, for the following sample.

   13 42 25 66 82 67 47 59 16 66
   56 23 66 78 66 23 12 99 87 55

2. Find the mean, median, and mode, for the following sample.

   113 422 235 66   822 667 947 579 106 274
   526 243 166 748 166 253 182 979 879 432

3. Use R to find the mean and median of the for the speed variable in the cars dataset.

4. Use R to find the mean and median of the for the decrease variable in the Orchard-Sprays dataset

Chapter 5

What are Measures of Dispersion?

In this chapter, you will learn the following...
- How to calculate values of dispersion
- How to assess the normality of a dataset
- How to make a box plot

Key Terms.
Range   Variance   Standard deviation
Kurtosis   Skew

Introduction

Measures of dispersion help you to understand how spread out the data is. Using this knowledge in combination with measures of central tendency can help you to better utilize information from the data. We will look at several ways to assess the dispersion of data in this chapter.

Range

The range is simply the difference between the highest and lowest value in a dataset.

Example 5.1: Find Range Using R

Find the range of the data below.
1 4 2 6 5 7 3 5 6 9
The highest value is 9 and the lowest is 1. Therefore,

$$Range = 9 - 1 = 8$$

Using R, we will find the range of the Sepal.Length variable in the iris dataset. We did this previously in the chapter on visualization. We need to use the range function as shown in the code below.

```
> range(iris$Sepal.Length)
[1] 4.3 7.9
```

We have the lowest and highest value in the variable but not the range. To do that, we need to subtract the highest value from the lowest as shown below.

```
> 7.9-4.3
[1] 3.6
```

3.6 is our range.

Variance & Standard Deviation

Variance and standard deviation are two highly related terms. They both deal with how data is spread out or disperse from the mean or from about the center of the dataset. In other words, the variance/standard deviation is the average distance a given data point is from the mean of the sample.

Let's use a visual to make sense of this. In Figure 5.1 below, there is a graph of the first ten values of the Sepal.Length variable in the iris dataset followed by a table with the values in them.

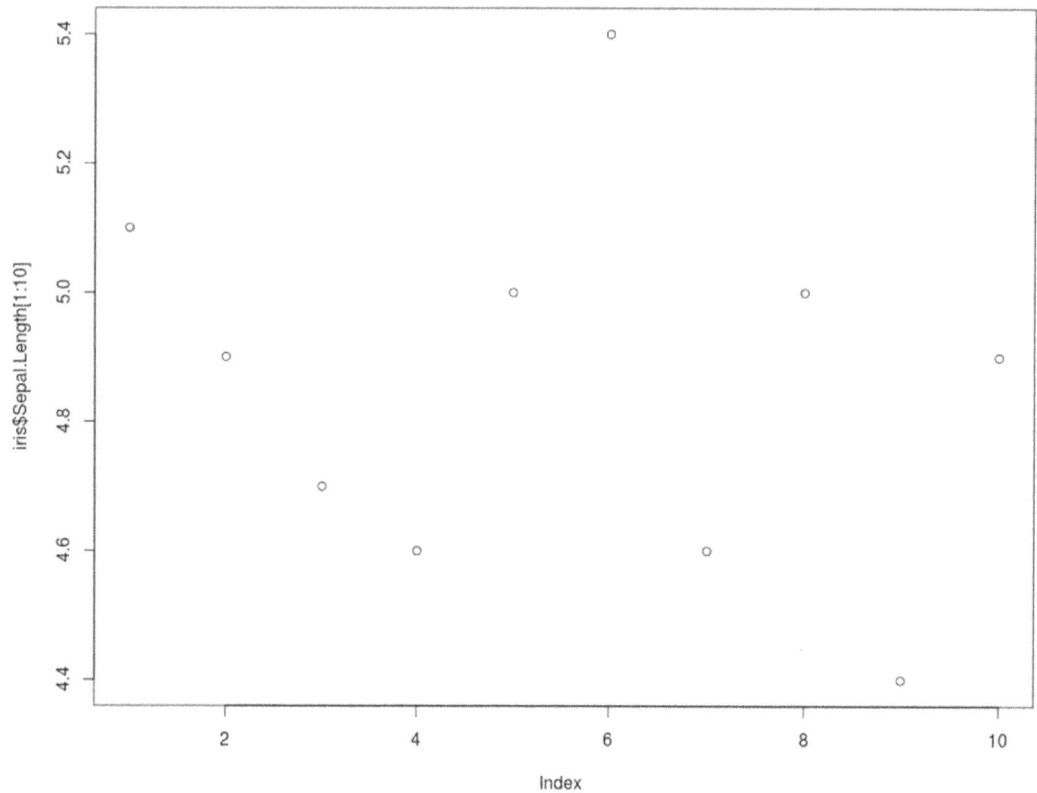

Figure 5.1: First Ten Data Points in Sepal.Length

```
##      mean
## 1    5.1
## 2    4.9
## 3    4.7
## 4    4.6
## 5    5.0
## 6    5.4
## 7    4.6
## 8    5.0
## 9    4.4
## 10   4.9
```

You can see the data in Figure 5.1 is spread out or disperse but right now we do not how spread out or disperse the data is. Right now we can tell it is spread out by using our eyes but we need numbers to confirm what we are seeing and make it objective.

To determine how spread out the data is we must compare each data point to some sort of standard. The standard we will use is the mean. In other words, we want to see how far each data point is from the mean because this is the standard we picked. In Figure 5.2, a line was drawn where the mean of Sepal.Length is. The mean is 4.86 and this value 4.86 is our standard in this dataset.

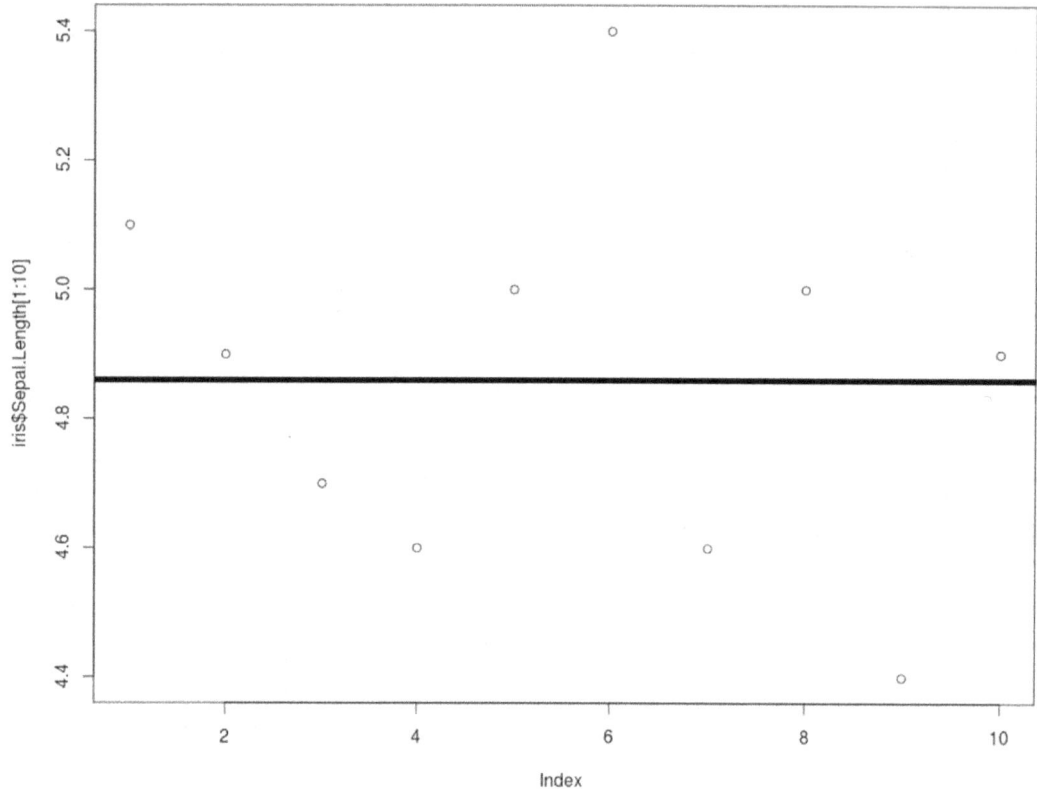

Figure 5.2: Sepal.Length Plot with Mean Line

As you can see, no observation is exactly on the line of the mean. Whenever an observation is not exactly the same value as the mean the difference between these two values is called error since we cannot explain why this data point is different from the mean or why it "deviates" from the "standard." The "standard" in standard deviation is the mean and the "deviation" in standard deviation is the average distance a point is from the mean. In Figure 5.3 the dotted line represents the error of each observation.

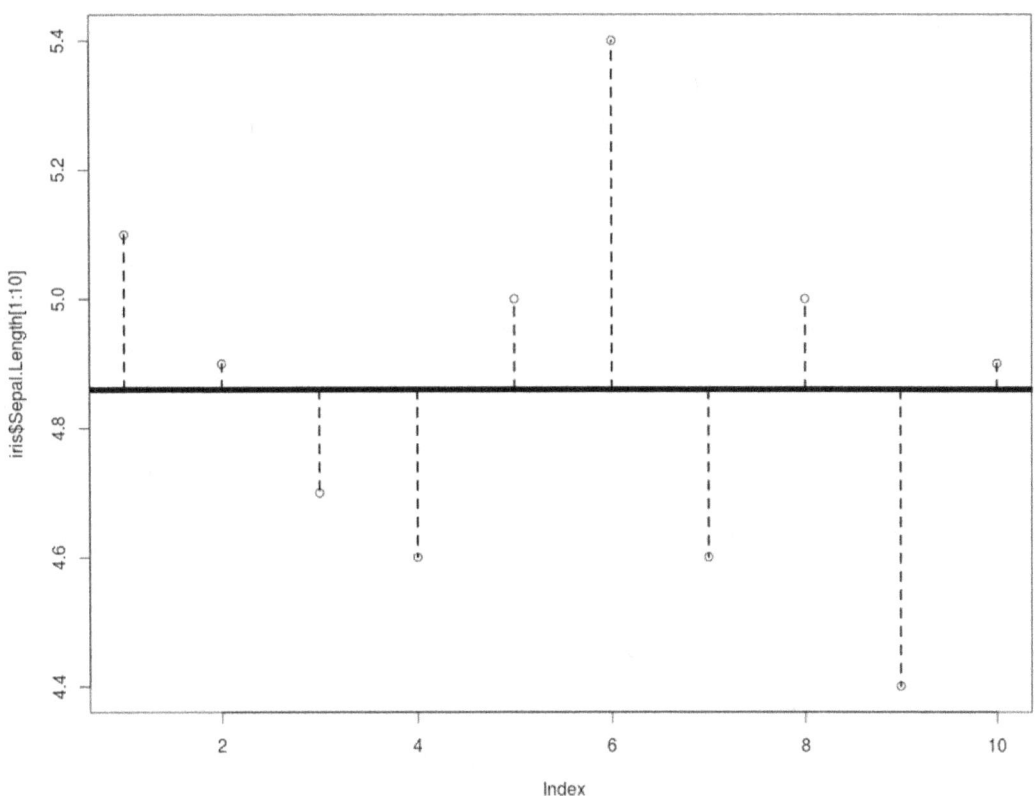

Figure 5.3: Sepal.Length with Error

You can tell that there is different amounts of error for each observation. In Figure 5.4 the distance of each point from the mean is shown and we also include this in a table. The diff column is the individual data point subtracted from the mean.

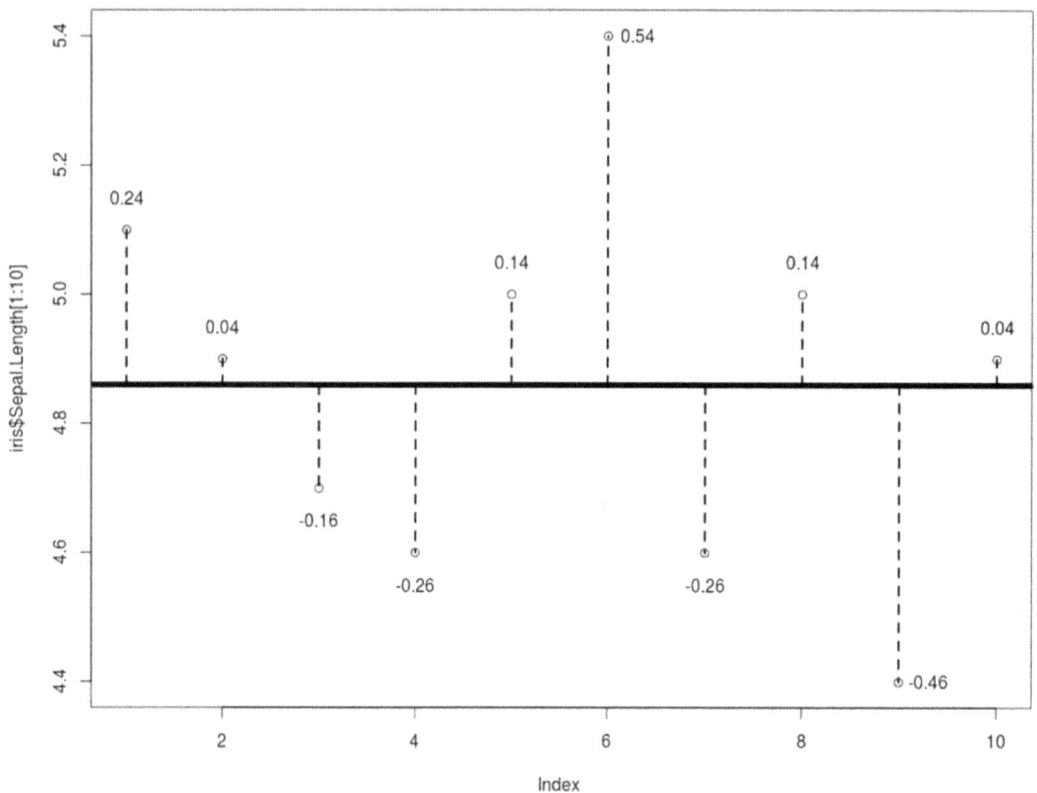

Figure 5.4: Error Amount for Each Observation of Sepal.Length

```
##      X  diff
## 1  5.1  0.24
## 2  4.9  0.04
## 3  4.7 -0.16
## 4  4.6 -0.26
## 5  5.0  0.14
## 6  5.4  0.54
## 7  4.6 -0.26
## 8  5.0  0.14
## 9  4.4 -0.46
## 10 4.9  0.04
```

To find the variance and finally the standard deviation we need to square the difference. We do this because if we add the numbers together now the positive and negative values will cancel each other and we will have a sum close to or even zero. You can check this yourself. By squaring the numbers they all become positive which is what we need. Below are the results.

```
##      X    diff square
## 1   5.1   0.24 0.0576
## 2   4.9   0.04 0.0016
## 3   4.7  -0.16 0.0256
## 4   4.6  -0.26 0.0676
## 5   5.0   0.14 0.0196
## 6   5.4   0.54 0.2916
## 7   4.6  -0.26 0.0676
## 8   5.0   0.14 0.0196
## 9   4.4  -0.46 0.2116
## 10  4.9   0.04 0.0016
```

If we add all the values together in the square column we get 0.76. To finally get the answer we need to divide 0.76 by 9. We divide by 9 because this is one less than the total observations of ten in our dataset. Below is the equation.

$$\frac{0.76}{10-1} = 0.084$$

Our variance is 0.084. The equation below is how to calculate variance. We just did this before knowing the equation.

$$variance = s^2 = \frac{\Sigma(x-\bar{x})^2}{n-1}$$

Often, the variance has little value for us by itself. Instead, we use the standard deviation. The standard deviation is simply the square root of the variance. Therefore,

$$\sqrt{0.084} = 0.291$$

Our standard deviation is 0.29. In other words, the average amount of error or departure from the mean is 0.29 in our example. What this means is that if you pick any random observation from are sample that observation will be within 0.29 units above or below the mean. The equation for standard deviation is shown below.

$$standard\ deviation = s = \sqrt{\frac{\Sigma(x-\bar{x})^2}{n-1}}$$

CHAPTER 5. WHAT ARE MEASURES OF DISPERSION?

Now that we have gone through an example let's do one example together manually and one using R.

The steps for finding the standard deviation are as follows

1. Find the mean

2. Subtract the mean from each x value

3. Square the difference of x and the mean

4. Find the variance

5. Square root the variance to find the standard deviation

Example 5.1

Find the standard deviation of the quiz scores below.

X
80
40
20
60
50
70
30
50
60
90

Step 1: Find the mean

$$\frac{80+40+20+60+50+70+30+50+60+90}{10} = 55$$

Step 2: Subtract the mean from each value of X

X	$X - \bar{X}$
80	80 - 55 = 25
40	40 - 55 = -15
20	20 - 55 = -35
60	60 - 55 = 5
50	50 - 55 = -5
70	70 - 55 = 15
30	30 - 55 = -25
50	50 - 55 = -5
60	60 - 55 = 5
90	90 - 55 = 35

Step 3: Square the difference

X	$X - \bar{X}$	$(x - \bar{X})^2$
80	80 - 55 = 25	625
40	40 - 55 = -15	225
20	20 - 55 = -35	1225
60	60 - 55 = 5	25
50	50 - 55 = -5	25
70	70 - 55 = 15	225
30	30 - 55 = -25	625
50	50 - 55 = -5	25
60	60 - 55 = 5	25
90	90 - 55 = 35	1225

Step 4: Find the variance

$$\frac{625+225+1225+25+25+225+625+25+25+1225}{10-1}$$

$$\frac{4250}{9} = 472.22$$

Step 5: Square Root the number to find the standard deviation

$$\sqrt{472.22} = 21.73$$

The standard deviation of the quizzes is 21.73. This means that the average quiz was about 22 points from the mean of 55.

CHAPTER 5. WHAT ARE MEASURES OF DISPERSION?

Find Standard Deviation Using R

Doing this in R is much easier. If I want to find the standard deviation of the Sepal.Length variable in the iris dataset, I only have to use the sd function to calculate the value.

```
> sd(iris$Sepal.Length)
[1] 0.8280661
```

Standard deviation can also be thought of as units. For example, for the R example the standard deviation was 0.82. If we want to figure out the value of two standard deviations we simply multiply 0.82 by 2 and get 1.64 to determine 3 standard deviations is 2.46. You can always figure this out by multiplying by 1, 2, or 3. This is important because most data falls within 1 standard deviation of the mean. In other words most data is the

$$\text{Mean} \pm \text{One standard deviation}$$

Therefore, in our iris example. Most values are between the two points below

$$5.84 - 0.82 = 5.02$$
$$\text{or}$$
$$5.84 + 0.82 = 6.66$$

Most data is between 5.02 and 6.66 for Sepal.Length in the iris dataset. This will make more sense in the future. Just for now, it is important to know that the standard deviation is a critical statistic. In fact it is used for calculating the following other statistical information.

Kurtosis	Skew	z-score
Confidence intervals	z-test	t-test

Quartiles

There are times when as a researcher you may need to split your dataset into equal sized groups, which can be used as a categorical variable. For example, it is common to split a dataset into four equal sized groups. If the data is split into four groups each group would contain 25% of the data. This is also called quartiles.

There are other ways to split data such as into 100 groups, which we also call percentile. However, quartile is probably more important for us because they are used with box plots as we shall see latter. The dataset below is divided into four parts. Notice how there are three quartiles that makes four parts. The quartile line is drawn it 5, 9, and 13.

```
  1  2  3  4  5    6  7  8  9    10  11  12  13   14  15  16  17
              1st              2nd                3rd
```
Below is the equation for determining the quartile.

$$Q_k = \frac{k(N+1)}{4}$$

Q_k = Quartile
N = Population
K = quartile location
Let's go through an example

Example 5.2

Below are the test results for a statistics class. Find the quartiles.
   10 11 25 25 37 44 49 49 58 64 75 77 86 88 91 97 100

To complete this problem we need to do the following steps.

Step 1: Arrange the data in order
Step 2: Find the quartiles
Step 3: Indicate where the quartiles are
Step 4: Arrange the data in order

Step 1: Arrange the data in order

This was already done for us.

Step 2: Find the quartiles

$$Q_1 = \frac{1(17+1)}{4} = 4.5 \approx 5$$

Our first quartile is at the fifth number in the dataset or 37. Lets find the other two quartiles.

$$Q_2 = \frac{2(17+1)}{4} = 9$$

$$Q_3 = \frac{3(17+1)}{4} = 13.5 \approx 14$$

Step 3: Find actual quartiles

Below is the data with the filled in quartiles

10 11 25 25 37 44 49 49 58 64 75 77 86 88 91 97 100
            1st   2nd   3rd

Notice how the actual values do not really matter when trying to find the quartiles. Rather it is the rank of the value that matters and not the value itself.

Finding Quartiles Using R

Below is an example using R code to find the quartiles of the Sepal.Length variable in the iris dataset. We will use the summary function to calculate this.

```
> summary(iris$Sepal.Length)
   Min. 1st Qu.  Median    Mean 3rd Qu.    Max.
  4.300   5.100   5.800   5.843   6.400   7.900
```

Notice how the summary function gives information you do not need in this situation such as the mean, minimum, and maximum variables. You may have also noticed how R uses the term median in place of 2nd quartile. This is because the median and the second quartile are the same value.

Box Plots

Box plots are a visual extension of quartiles. It's easier to explain box plots by first looking at them so Figure 5.5 is an example.

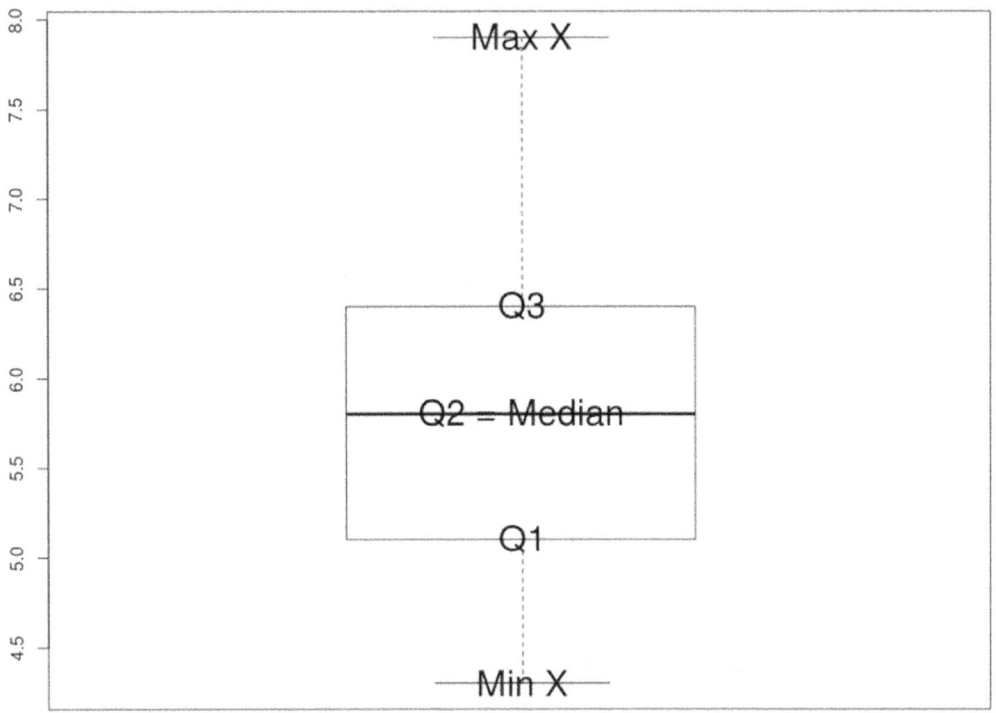

Figure 5.5: Example Box Plot

In Figure 5.5, you can see that the tail of the box plot is the lowest X value, Q1 represents the 1st quartile. The line in the middle of the box represents 2nd quartile or median. The top of the box represents the 3rd quartile and the top tail represents the highest value of X.

In Figure 5.6 is the same plot as the one above but the values for the minimum X, Q1, Q2, Q3, and Max X are added.

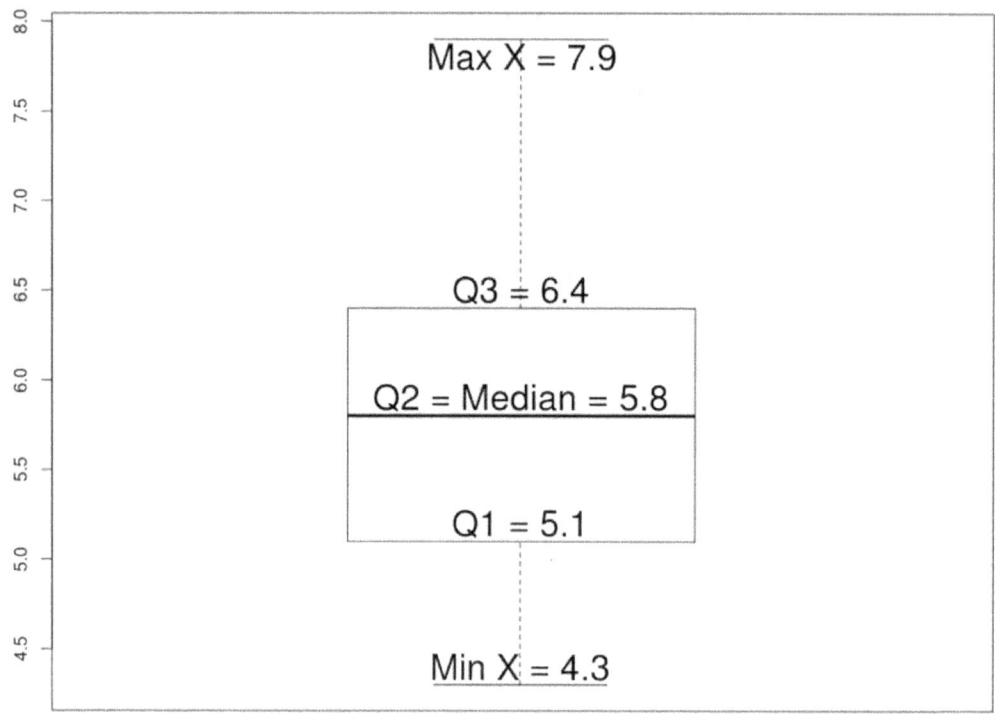

Figure 5.6: Example Box Plot with Values

This is the same information we get using the summary function as show below.

```
> summary(iris$Sepal.Length)
   Min. 1st Qu.  Median    Mean 3rd Qu.    Max.
  4.300   5.100   5.800   5.843   6.400   7.900
```

Make Box Plot Using R

You never make box plots by hand so we are going to make a box plot of the Petal.Width variable in the iris dataset. The code is as follows. Figure 5.7 is the box plot.

```
boxplot(iris$Petal.Width)
```

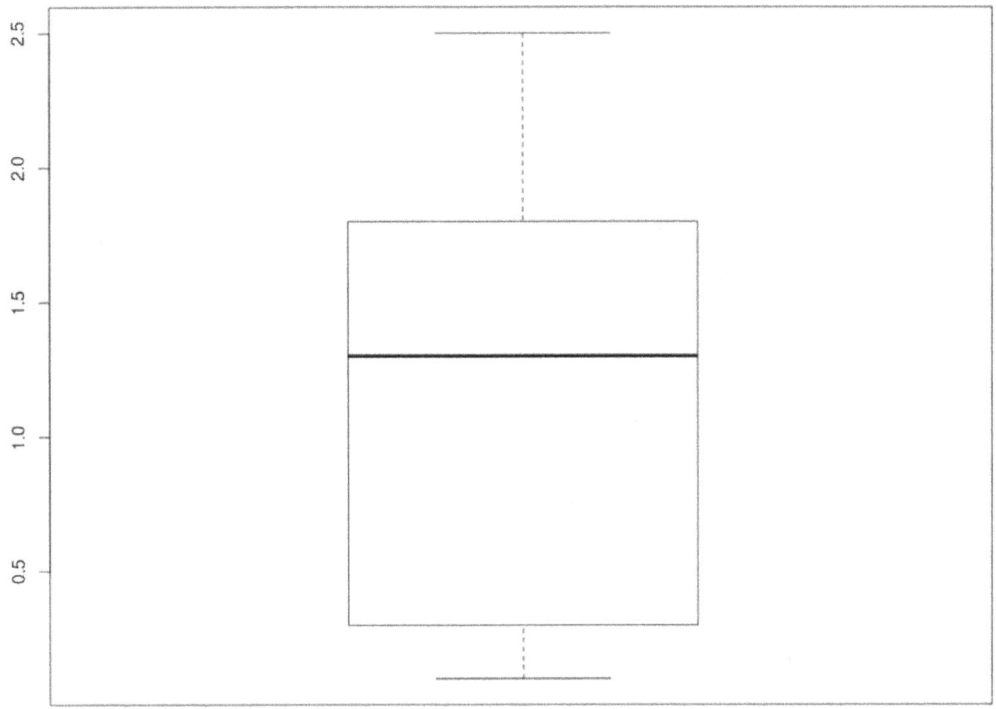

Figure 5.7: Petal.Width Box Plot

The summary function provides the details for the Petal.Width variable.

```
> summary(iris$Petal.Width)
   Min. 1st Qu.  Median    Mean 3rd Qu.    Max.
  0.100   0.300   1.300   1.199   1.800   2.500
```

Kurtosis

Kurtosis is a measure of how peak or flat a distribution is. When we talk of peak or flat it must be in comparison to something. Just as with standard deviation in which every data point is compared to the mean kurtosis involves a comparison. When we look at a distribution, we compare it to some other distribution to determine if it is reasonable. The most commonly used distribution for comparison is the normal distribution. We will talk about this more in a future chapter. Right now, you need to know that kurtosis compares your datasets peak or flatness to what is expected of a normal distribution. Examine the graph below in Figure 5.8. Distribution A is normal while distribution B and C are not,

As you can see distribution A is balance and symmetrical or in other words normal. Distribution B is really tall and skinny while distribution C is flat and fat if you will. Distribution A is known as mesokurtic, B as Leptokurtic, and C as platykurtic.

It is possible to calculate the amount of kurtosis in a distribution. For interpretation, any positive value indicates that the distribution is leptyokurtic, any negative value indicates the distribution is platykurtic, and a value of zero indicates a distribution is mesokuritc or normal.

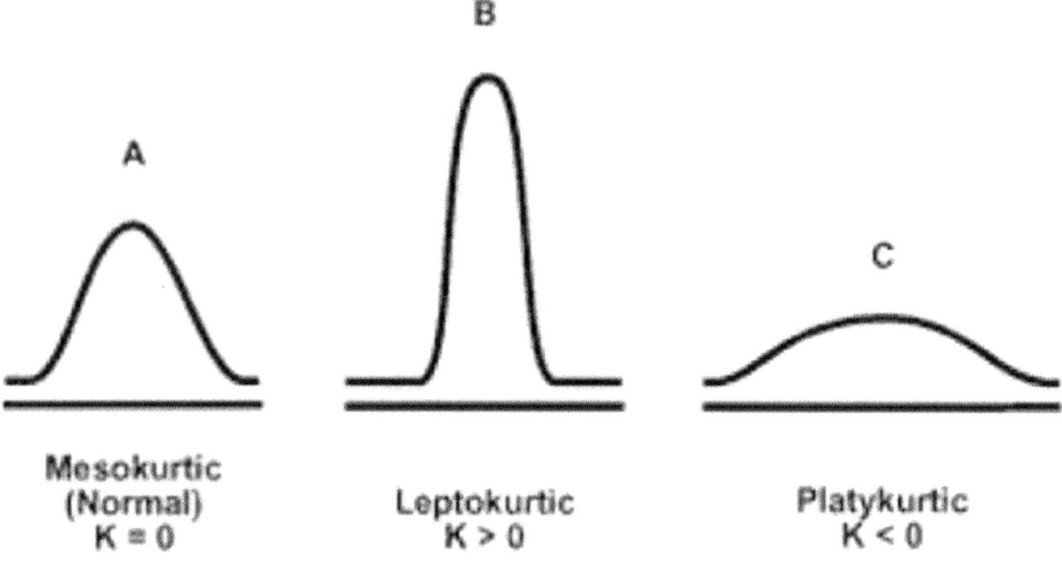

Figure 5.8: Different Distributions by kurtosis

Type	Value
Kurtosis > 0	Leptokurtic
Kurtosis < 0	Platykurtic
Kurtosis = 0	Mesokurtic

Table 5.1: Kurtosis Values

Calculating kurtosis by hand is somewhat complex. As such, we will just use R to find the kurtosis of the Sepal.Length variable in the iris dataset.

Find Kurtosis Using R

Example 5.3
To determine kurtosis you will need the `moments` package. You might need to download it using the `install.package` function. Below is the actually code

```
> library(moments)
> kurtosis(iris$Sepal.Length)
[1] 2.426432
```

With the positive value, you can see that are distribution is leptokurtic. However, it is important to note that unless the values are extreme it is difficult to see kurtosis problems in a plot. This is one reason for the statistical test. In addition, how much kurtosis is too much is not agreed upon among statisticians.

Skewness

Skewness is a measure of the shape of a distribution. A symmetrical distribution has the perfect bell shape and the mean, median, and mode are aligned. Positively skewed distribution has a long tail to the right and the mean is a higher value than the median with the mode being a smaller value than the median. Negatively skewed has a long tail to the left with the median now a higher value than the mean and the mode having the highest value. A visual of skewness is provided below in Figure 5.9.

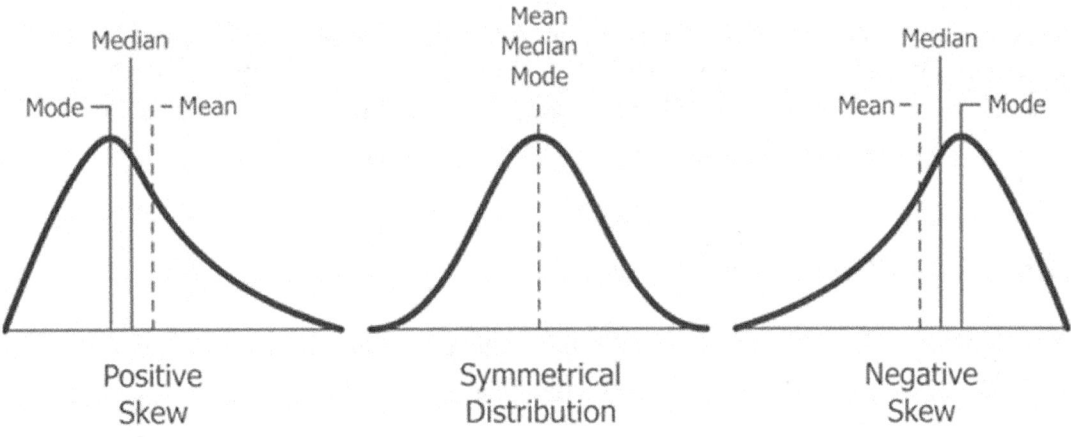

Figure 5.9: Different Distributions by Skewness

Find Skewness Using R

The calculation for skewness is not as complex as kurtosis but we can find it in R quickly for the Sepal.Length variable in the iris package using the code below.

```
> skewness(iris$Sepal.Length)
[1] 0.3117531
```

The results indicate that our distribution is slightly positively skewed which means there is a long tail to the right, the mean is a higher value than the median, and the mode is the lowest value of these three numbers. We can confirm the mean is greater than the median by using the summary function.

```
> summary(iris$Sepal.Length)
   Min. 1st Qu.  Median    Mean 3rd Qu.    Max.
  4.300   5.100   5.800   5.843   6.400   7.900
```

Both kurtosis and skew compare your data to an ideal normal distribution. This is done because many more advanced statistical test assume that your data is normally distributed. Understanding kurtosis and skew can help you to know if your data meets this assumption or if you need to make adjustments to your data analysis.

Conclusion

Measuring the dispersion of data is a critical component of many statistical tools. Perhaps the most important measure of dispersion is standard deviation. The results of standard deviation are used in many calculations. Box plots provide a powerful visual of the spread of the data as well. There are also measures that indicate the normality of the distribution such as kurtosis and skew.

Points to Remember

- Standard deviation is one of the most important statistical tools as it is a foundational piece of information in many calculations.

- Box plots provide a visual of quartile information

- Skew and kurtosis are useful for assessing normality of a distribution

R Code Used

- sd(): Calculates standard deviation
- summary(): Provides summary statistics
- boxplot(): Makes box plot of the data
- kurtosis(): Shares the kurtosis of the object
- skewness(): Shares the skewness of the object

Exercises

1. Below are exam scores from a course. Calculate the variance and standard deviation.

X	$X - \bar{X}$	$(X - \bar{X})^2$
78		
34		
89		
45		
88		
77		
70		

2. Below are the hours students work in a semester. Calculate the variance and standard deviation

X	$X - \bar{X}$	$(X - \bar{X})^2$
48		
34		
59		
45		
28		
67		
50		

3. Calculate the standard deviation of the speed variable in the cars dataset

4. Find the skew and kurtosis of the speed variable in the cars dataset

5. Make a box plot of the speed variable in the cars dataset.

Chapter 6

What is Probability?

In this chapter, you will learn the following...

- The role of probability in statistics

Key Terms.
Event   Sample Space   Potential Outcomes

Introduction

Probability is a concept that can be difficult to understand. The reason being is that probability is trying to measure or determine something that has not happen yet. Through looking at prior results probability can provide an indication of what to expect in the future. Therefore, understanding probability is critical to appreciating more advance statistical concepts such as normal distribution, hypothesis testing, and all of statistics going forward. This chapter only provides the most critical information needed to grasp the bare basics of probability.

Definition and Example

Probability is a measure of the likelihood that a specific event will occur. A probability of 0 means that the event cannot happen while a probability of 1 means that an event is certain. By event, we mean the result of a single trial. This may make better sense with an example.

One of the first examples many students learn about involving probability is flipping a coin. As you know, there are two possible outcomes for flipping a coin and these are heads and tails. To determine the probability of getting heads you need to consider the following.

$$\frac{\text{Number of possibilites that meet the condition}}{\text{Number of potential outcomes}}$$

To make things clear, heads and tails are the potential outcomes of our example. In other words, nothing else can happening except getting a head or tail. Every time we flip the coin this is know as an event. We can flip the coin one time or a million.

In addition, each number on the die has an equal likelihood to occur. What this means is that the probability of get 1 is the same as getting a 2 or some other number. If equal likelihood was not the case in this example we would have to adjust how we calculate the results.

In our current example with the coin we have one possibility which is heads while we have two potential outcomes which are heads or tails. Therefore, our probability of getting heads is as follows.

CHAPTER 6. WHAT IS PROBABILITY?

$$\frac{\text{Number of possibilites that meet the condition}}{\text{Number of possibly outcomes}} = \frac{1}{2} = 50\%$$

You can see that we expect that 50% of our flips should be heads. However, as we shall see, this is what we expect but sometimes our expectations do not match what we observed.

Another common example for explaining probability involves the use of a die. Let's say that you have a die. You already know that the probability of getting the number 2 is 1/6. The values 1-6 is the sample space or the list of potential outcomes. When we roll the dice and get a number, such as 3, this is an event. Below is the probability of rolling a any number 1 time with a die.

$$\frac{\text{Number of possibilites that meet the condition}}{\text{Number of possibly outcomes}} = \frac{1}{6} = 16.66\%$$

If you have been reading carefully you would know that a die can be considered as a nominal or ordinal variable. In other words, you cannot get the value 5.5 or 2.3 when rolling a die. The die has what we call are discrete values. The example of the coin flip and the die roll involve discrete probability. By discrete it means the same as the categories in a categorical variable.

Furthermore, you can roll a 1 but not a 1 and a 2 at the same time. In other words, the potential outcomes are mutually exclusive. This is a fancy way of saying that 2 or more outcomes cannot happen at the same time.

If there is an equal probability for each number on a die we should expect that if we rolled the die many times each number would happen about the same. Below in Figure 6.1 there is a histogram of 5000 rolls of a die. Notice how balance the values are.

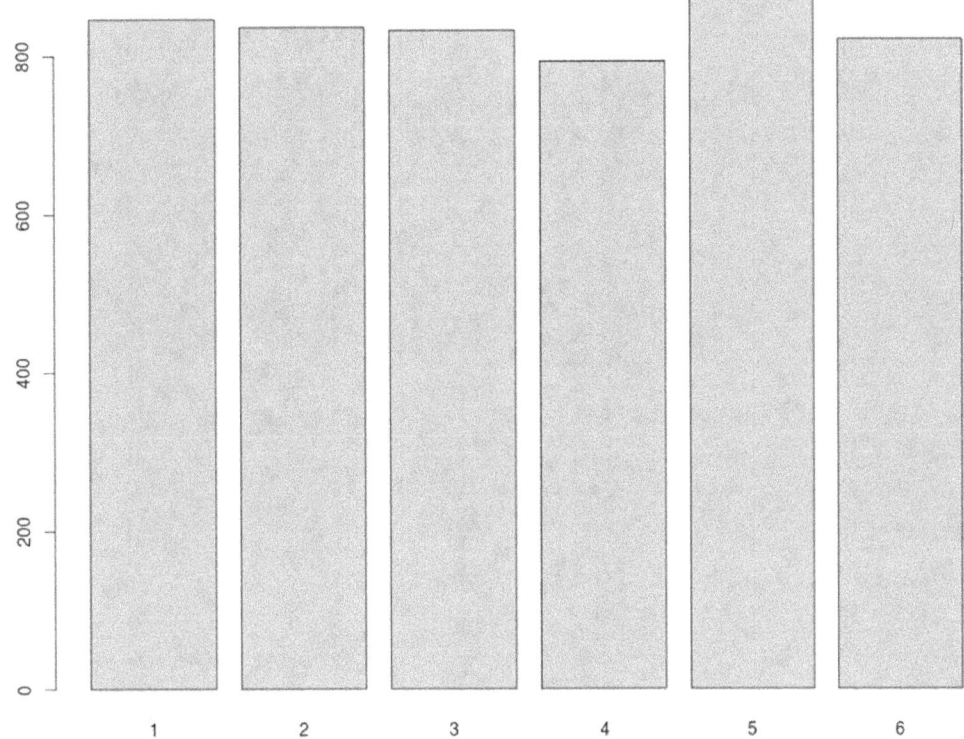

Figure 6.1: Results of Rolling a Die 5000 Times

A question you may be wandering is way are the bars so balanced? This is because of the equal probability of each event. This is also known as the law of large numbers. As the number of times you conduct the experiment, which in this example means rolling the die, it increases the stability of the results that you expect. If we had rolled the die only 5 times it would not be this balanced. However, over 5,000 it is exceedingly balanced.

It needs to be noted that when you are considering multiple events such as rolling the number 2 three times in a row that the probability changes. However, this is beyond the scope of this book.

Continuous Probability

When we look at probability from a sample or population for a continuous variable like body weight, we do not get equal probabilities for every value. Some values are more likely to happen than others are and we can use histograms to see this. Histograms do not tell you the frequency of certain values but can also be used to indicate the probability of certain range of values. For example, if you look at the histogram in Figure 6.2 of the Sepal.Length variable in the iris dataset you get the following.

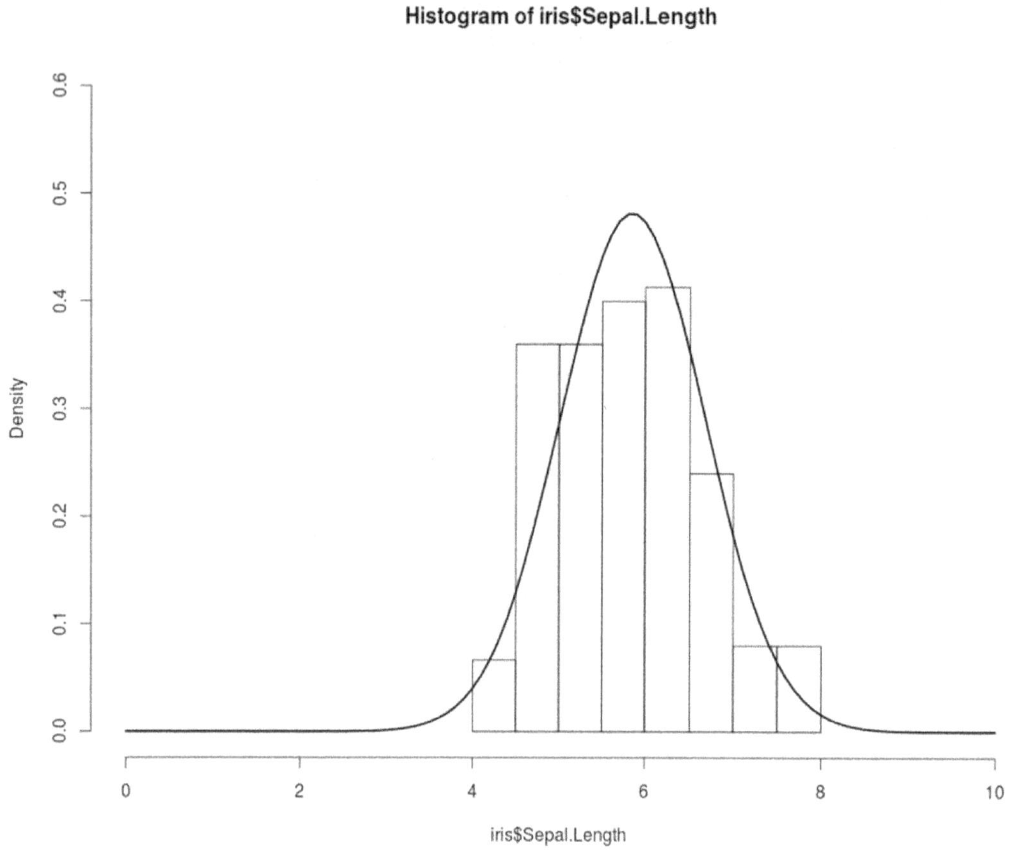

Histogram of iris$Sepal.Length

Figure 6.2: Histogram of Sepal.Length

In the histogram above, you can see that not all values are equally likely to occur. For example, values around 6 are the most likely to occur. In other words, it is most likely or probability that an iris will have a sepal length of about 6.

The curved line in the graph provides a visual of the probability we would expect if this data had a normal distribution (this will be covered later). As the line rises so to does the probability of the event happening. Values less than 4 and greater than 8 are rare and it is nearly impossible to obtain such values based on the data that we currently have. Remember that this is a sample of irises. This is why I said nearly impossible to obtain values less than 4 or more than 8. I am inferring or making a conclusion from my sample that this is almost impossible. By knowing the probability of a given event or value you can have an idea of what to expect. For example, I know that if I find an iris with a sepal length of 10 or greater that is really weird. I know this because of my knowledge of the probability of this happening from my sample. In other words, probability can help you to determine what is normal for your data.

Since our variable is continuous we can never determine specific or discrete values. This is because the probability of a specific value happening is zero. The reason for this is that there are an infinite number of possible values that can happen. For example, we can get 6 or 6.1 or 6.01 or 6.001 etc. Since the possibilities are infinite the probability of any given number is zero.

This is why with continuous probability we always think in terms of ranges rather than specific value. For example, the probability that an iris has a sepal length greater then 6 or less than 6. Each of these is a range of values and this can be calculated as we will see in a future chapter.

We are assuming some things as we look at the histogram above. One, we are assuming that all events are mutually exclusive which means that one event does not affect another. In other words, the sepal length of one flower does not affect the sepal length of another. Two, the events in the histogram are marginal probabilities, which mean they can happen without considering any other event.

I also need to mention that I am leaving out a lot of the technical information on probability, as this is confusing for people unfamiliar with statistics. Lastly, the iris example applies to a normal distribution but there are many other distributions such as Poisson. The difference between distributions is there shape while most of the other assumptions remain the same.

The idea of what we expect versus what we get relates with statistical significance and is a cornerstone of inferential statistics. Statistics is all about comparing what we observed versus what we expected and determining if there is a difference between the two. For now, this is all that will be mentioned about this but this idea will become highly important in a few more chapters.

Conclusion

You can take entire courses on probability and learn some interesting things. However, our purpose here is to know enough about probability to apply it to our use in statistics. You need to remember that probability is the likelihood of something happening. You also need to know that we can determine the actual probability of an event using statistics. In order to do this, you will need to have additional information about distributions and hypothesis and this is what we will examine in the near future.

Chapter 7

What is Normal Distribution?

In this chapter, you will learn the following...

- The use of the normal distribution
- The purpose of the standard normal distribution
- How to calculate z-scores

Key Terms.
  Normal distribution   Sample distribution   Standard normal distribution

Introduction

We have alluded to the normal distribution for several chapters now and are finally going to focus exclusively on it. It is in this chapter that we tie together ideas about standard deviation, skew, kurtosis, and probability. Understanding normal distribution prepares us for understanding hypothesis testing and confidence intervals.

When we use the term normal distribution we are really talking about normal continuous distribution. Remember that probability can involve discrete or continuous variables. This chapter is focused mostly on understanding distributions involving continuous data.

Understanding Normal Distribution

The normal distribution is bell-shaped distribution that is often found when values gather around the mean. Another word for normal is expected which means that when we collect data we anticipate that it will take the shape of the normal distribution in many situations. Generally, as the sample size increases the more bell-shaped or normal a distribution becomes. If you look at Figure 7.1, you can see that as the sample size gets larger the bell-shape becomes more and more balanced or symmetrical.

This idea that a distribution becomes more and more normal as the sample size increases is known as the central limit theorem. It is so important that many statistical tests assume a normal distribution.

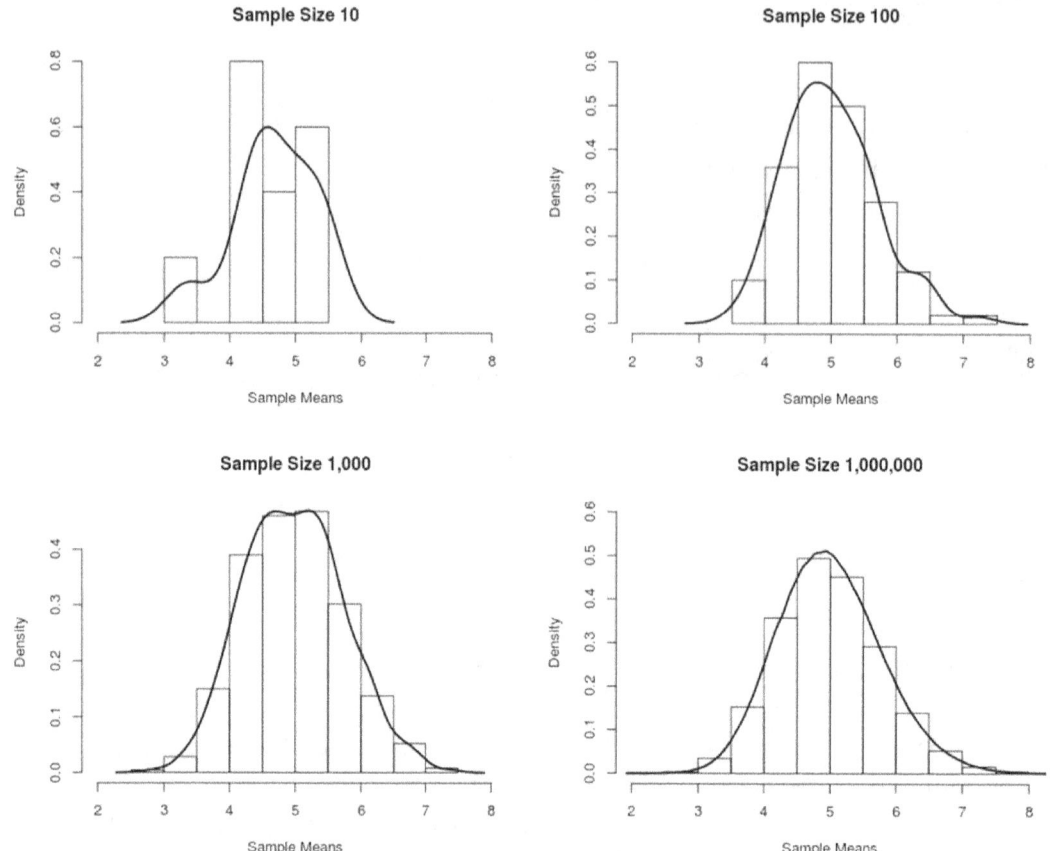

Figure 7.1: Plots of Different Sample Sizes
In order for a distribution to be normal it must meet the following characteristics.

- Bell-shape distribution

- Mean, median, and the mode are at the center of the distribution

- Unimodal

- Continuous variable

- Area under the curve (under the bell) must equal to 1.00 or 100%

Few sample distributions meet all of these characteristics perfectly. However, it is common for a sample distribution to closely meet these characteristics so that the normal distribution can be used as an approximate model for different statistical test.
The normal distribution can also be thought of in terms of standard deviations. In a previous chapter, we spoke about how the majority of the values of a distribution are within 1 standard deviation of the mean. Actually, there is a 68% probability that the

values in a normal distribution are within 1 standard deviation of the mean. This means that most values are within 1 standard deviation above the mean or 1 standard deviation below it. For two standard deviations, the probability improves to 95% and with 3 standard deviations the probability improves to 99%. If you recall from example 5.1 we had ten quiz scores.

X
80
40
20
60
50
70
30
50
60
90

The mean of these scores was 55 and the standard deviation was 21.73. What this means is as follows

1 standard deviation = 21.73
2 standard deviations = 43.46 (Multiply the standard deviation by 2)
3 standard deviations = 65.19 (Multiply the standard deviation by 3)

Now our sample size here is very small so this is not a prefect approximation but below is what we are to expect

- 68% of the values are within one standard deviation above or below the mean 33.27-76.73. This answer was calculated by subtracting 21.73 (one standard deviation) from the mean of 55 for the smaller number and adding 21.73 to 55 for the larger number.

- 95% of the values are within two standard deviation above or below the mean 11.54-98.46. This answer was calculated by subtracting 43.46 (two standard deviation) from the mean of 55 for the smaller number and adding 43.46 to 55 for the larger number.

- 99% of the values are within three standard deviation above or below the mean -10.19-120.19. This answer was calculated by subtracting 65.19 (three standard deviation) from the mean of 55 for the smaller number and adding 65.19 to 55 for the larger number.

CHAPTER 7. WHAT IS NORMAL DISTRIBUTION?

Because our sample size is so small the 3rd standard deviation does not make sense. This is because you cannot have a negative quiz score. In other words, our variable quiz score is a ratio variable. In Figure 7.2 is a histogram of our quiz score data. Notice how funny-shaped it is when compared to the "Sample Size 1,000,000" plot in Figure 7.1.

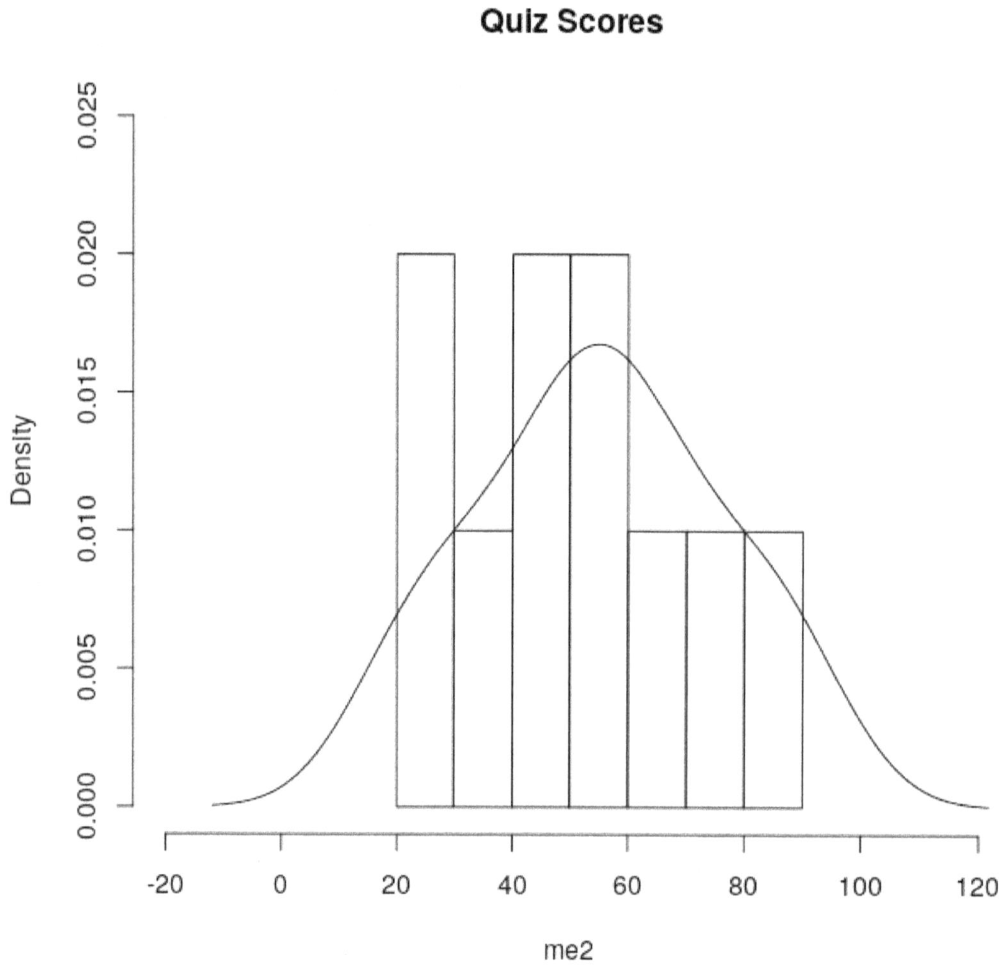

Quiz Scores

Figure 7.2: Quiz Score Histogram

Table 7.2 provides a summary of the quiz score distribution.

	Quiz Scores	Total Number	Percent
1st Standard Deviation	40,50,50,60,60,70	6	60%
2nd Standard Deviation	20,30,40,50,50,60,60,70,80,90	10	100%

Table 7.1: Quiz Distribution

Notice in Table 7.2, how 60% of the data is within one standard deviation and that 100% is within two standard deviations. Again, this is not perfect or completely normally distributed because the sample size is so small. However, you can all ready see how data is distributed. Items near the mean have a higher probability of occurring than items further from the mean.

Probability may still sound strange but remember that this is simply the likelihood of an event taking place. Keep in mind also that as we move further from the center of the normal distribution that those events further and further from the center are less and less likely to happen. This is the relationship between the normal distribution and probability. High probability events are near the center while less probably events are far from the center of a normal distribution. Figure 7.3 provides a breakdown of the normal distribution with the appropriate probabilities.

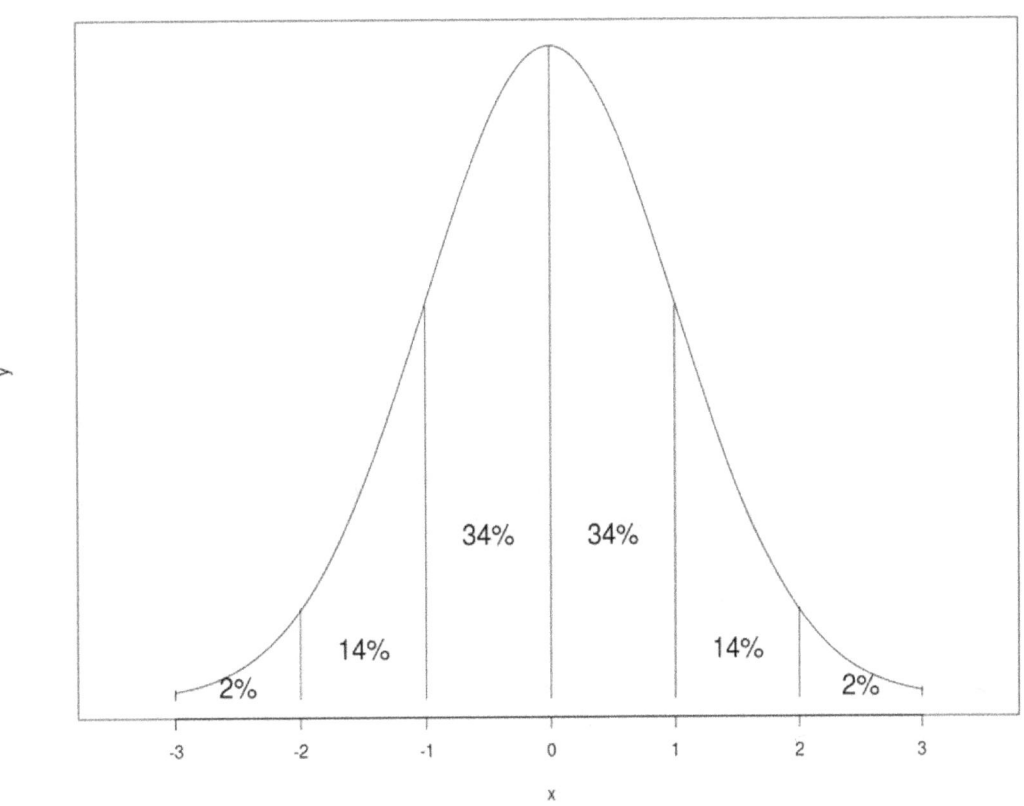

Figure 7.3: Normal Distribution and Probability

The x-axis has the mean at zero and the standard deviations from -3 to 3. As an example, from the mean of zero to the 1st standard deviation are 34% of the value. From the mean of 0 to 1 standard deviation below the mean (-1) is also 34%. Therefore, 68% of the

CHAPTER 7. WHAT IS NORMAL DISTRIBUTION?

values are within 1 standard deviation of the mean (34 + 34). This concept of data being dispersed almost totally within the first 3 standard deviations is known as the empirical rule. In addition, if you add up all the values under the curve it comes to 100% or 1.

$$2\% + 14\% + 34\% + 34\% + 14\% + 2\% = 100\%$$

The current example is somewhat abstract so we are going to look at the Sepal.Length distribution of the iris dataset to provide a realistic application. Figure 7.4 is the distribution of the Sepal.Length data in the iris dataset.

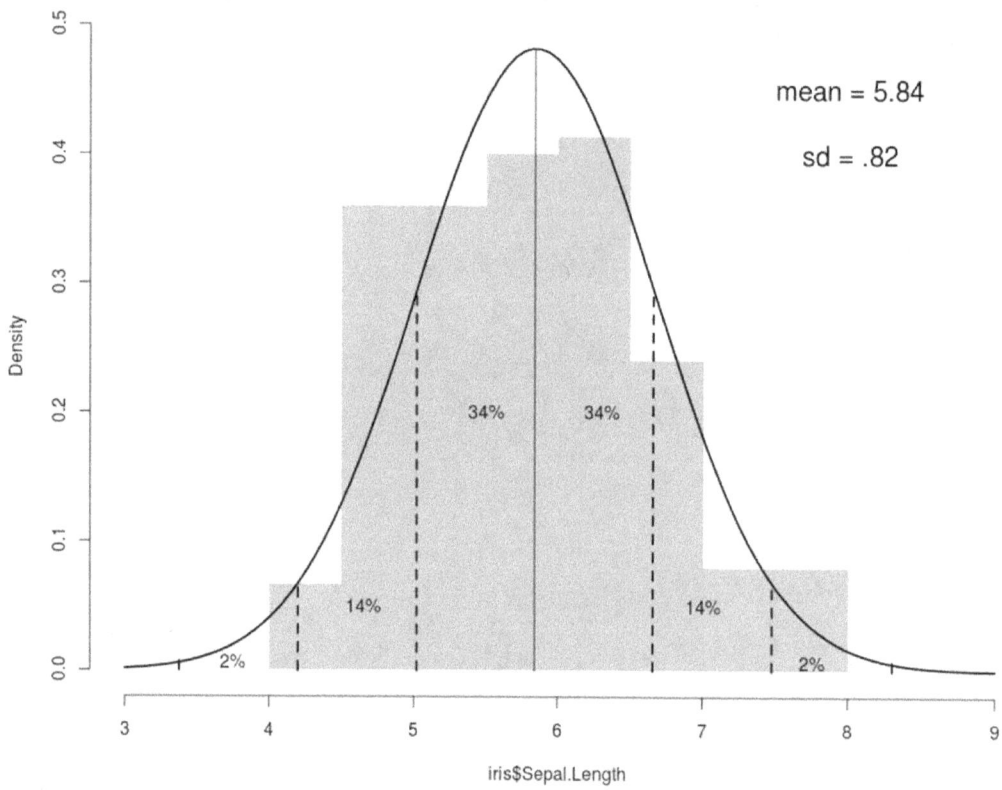

Figure 7.4: Sepal.Length Distribution with Probability

The solid line in the middle is the mean. The dash lines represent the different standard deviations. Table 7.2 provides the ranges for 1st, 2nd, 3rd, standard deviations.

Standard Deviation	Below mean	Above mean	Range
1 (68%)	5.84 - .82 = 5.02	5.84 + .82 = 6.66	5.02 - 6.66
2 (95%)	5.84 - $(.82^2)$ = 4.20	5.84 + $(.82^2)$ = 7.48	4.20 - 7.48
3 (99%)	5.84 - $(.82^3)$ = 3.38	5.84 + $(.82^3)$ = 8.30	3.38 - 8.30

Table 7.2: Confidence Interval Table

Most values in the histogram fall within 5.02 - 6.66. You can clearly see this. Almost all values are with 4.2 - 7.48. Understanding this is key to appreciating hypothesis testing, which we will discuss in a future chapter.

If you look closely you will see that our data in Figure 7.4 is not completely normal. Wherever the data does not fit the curve there are problems. For example, we have too many values around 4.8 and above 7.5. The point to remember is that most data is never normal when doing research. However, the application of the Normal Distribution still provides useful information for decision-making.

We need to cover one more point. There is a relationship between probability and percentile. For example, if you look at Figure 7.4 you can see how the distribution is divided. If I wanted to know what sepal length was in the 99% percentile I would look for any sepal length greater than about 8. In other words, the normal distribution can also be used to determine percentile rank.

Standard Normal Distribution

Understanding normal distribution is linked to probability. We can now determine the probability for a range of values by using the normal distribution. The question now is how do we do this?

To find the probability of any given observation we must use the standard normal distribution. The standard normal sets the mean to zero and the standard deviation to one of a normal distribution. This allows us to compare values to the same standard rather than constantly changing the mean and standard deviation of the distribution for every variable or dataset we want to examine. Figure 7.3 is actually an image of a standard normal distribution.

The standard normal distribution allows you to calculate something that we call the z value. The equation is below.

$$z = \frac{x - \mu}{\sigma}$$

z = z value
X = individual value
μ = mean of the distribution
σ = standard deviation

This value tells you the distance of a given x value from the mean in standardized units. The z-value tells you how many standard deviations a value is above or below the

CHAPTER 7. WHAT IS NORMAL DISTRIBUTION?

mean. It is presented as a range because the probability of a specific value is zero as mentioned in the previous chapter.

When using the standard normal distribution, once you get the z-score you have to look at a standardized normal distribution table to find the actually area of the curve that you want. Appendix A contains this information. It is best to look at a few examples to make sense of this.

Example 7.1

Find the area under the standard normal distribution curve between z = 0 & z = 1.75.

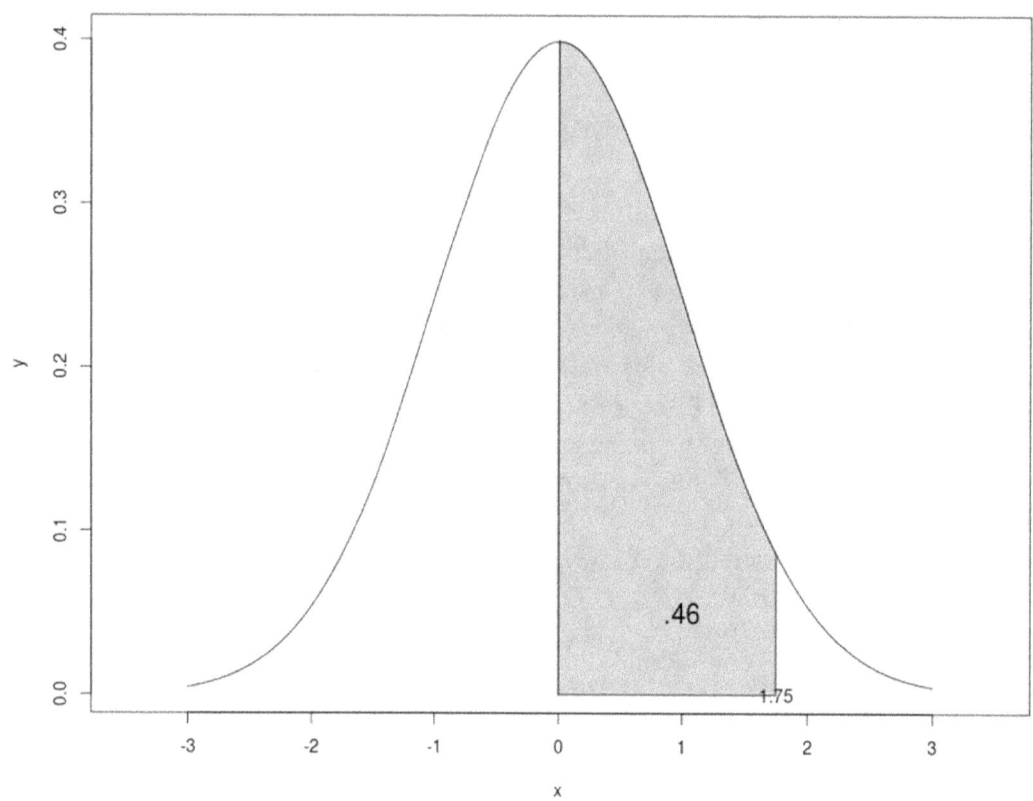

Example 7.1
To solve this we do the following.

1. Convert the z-scores to probabilities using the normal distribution table

2. Take these values and do basic subtraction.

The answer is found below

$$P(0 < z < 1.75) = .5 - .04006 = .46$$

In simple English, the area between z = 0 and z = 1.75 is .46 or 46% of the curve. We subtract 0.5 from the value from the table because we want the total distance from 0 to 1.75. Another way to explain this is that 46% of the values in the distribution are between 0 to 1.75 standard deviations above the mean.

You can also solve this using R by using the pnorm function. Below is the code.

```
> pnorm(1.75)-pnorm(0)
[1] 0.4599
```

Example 7.2

Find the area under the standard normal distribution curve between z = 0 & z = -2.20. To solve this we do the following

$$P(0 < z < -2.20) = .5 - .0139 = .48$$

In simple English, the area between z = 0 and z = -2.20 is .48 or 48% of the curve. or you could say 48% of the values are between 0 standard and 2.2 standard below the mean.
Below is the solution using R.

```
> pnorm(0)-pnorm(-2.20)
[1] 0.4861
```

CHAPTER 7. WHAT IS NORMAL DISTRIBUTION?

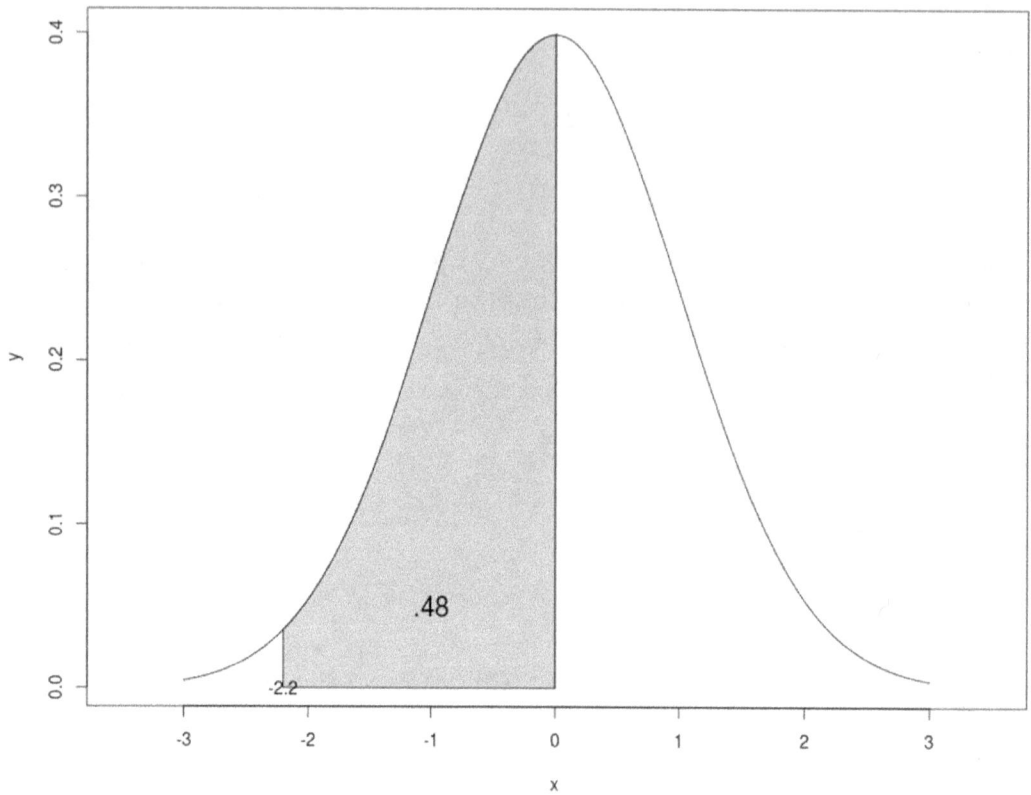

Example 7.2

Example 7.3

Find the area under the curve to right of z = 1.96
To solve this we do the following P(0 < z < 1.96) = 1 -.97 = .03 In simple English, the area between z = 0 and z = 1.96 is .03 or 3% of the curve. First we subtract 1 instead of 0.5 because we only want all the area to the right of 1.96.

Below is the r code

```
pnorm(0)-pnorm(1.96)+.5
[1] 0.025
```

We had to add 0.5 because we want everything from the unshaded region on not just from the middle line of 0.5.

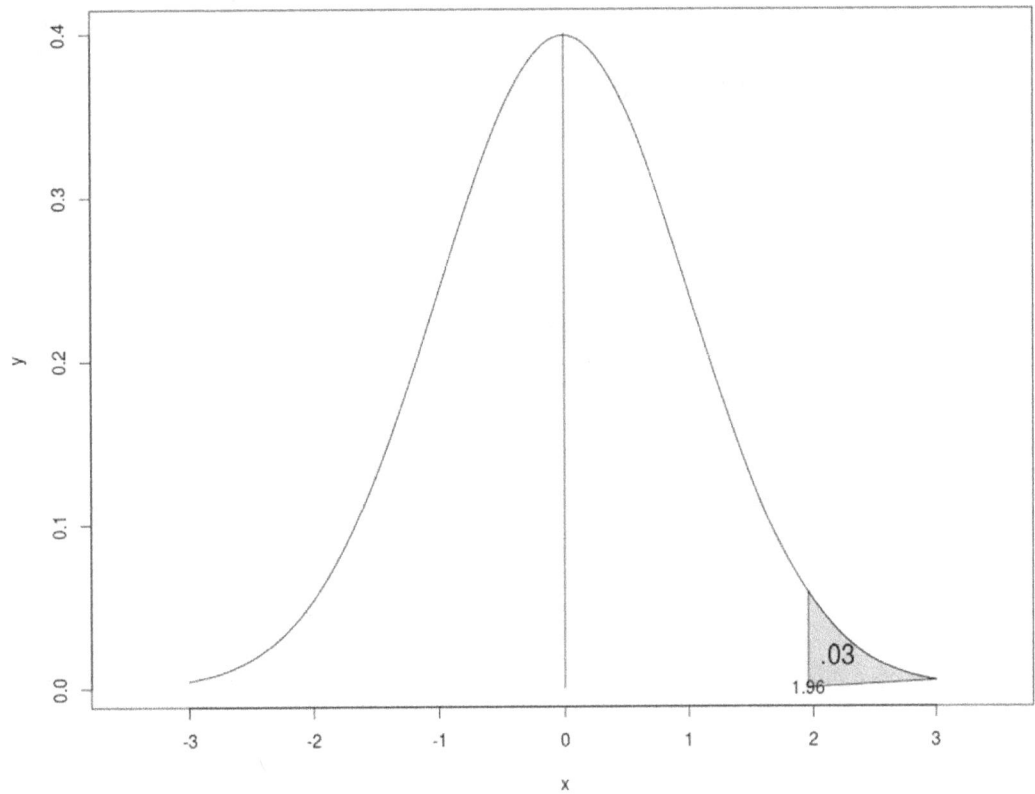

Example 7.3
Let's do an example using the Sepal.Length variable from the iris dataset.

Example 7.4

What is the probability that the sepal length of an iris is greater than 7.2. We simply need to do follow the equation
Mean = 5.84
Standard deviation = 0.82

$$z = \frac{X-\mu}{\sigma} = \frac{7.2-5.84}{.82} = \frac{1.36}{.82} = 1.65$$

We know our z = 1.65 we can now use our standardized table to find they are

$$P(X > 7.2) = P(z > 1.65) = 1 - .95 = .05$$

The probability that a sepal length is 7.2 or greater is 5%, Below is a diagram of example 7.4. The black area represents the sepal lengths greater than 7.2. You can see that this is a small region of the histogram.

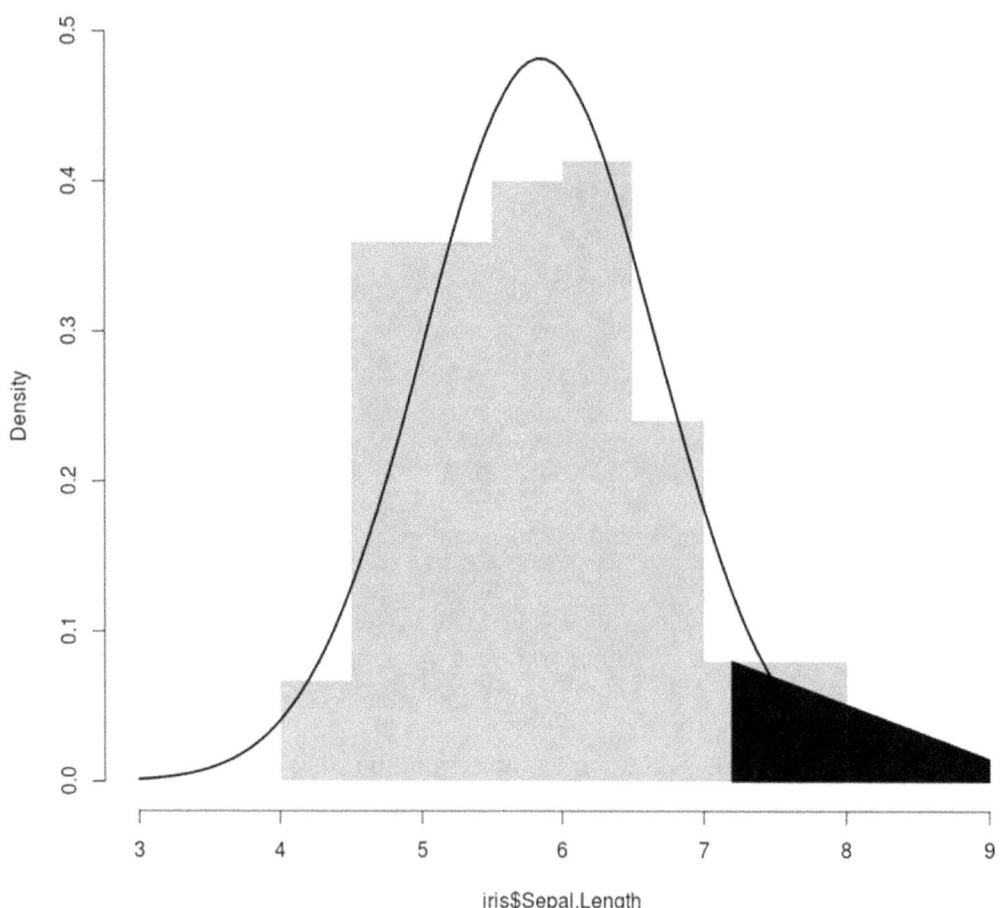

Example 7.4
It is not necessary to master these concepts right now. This information is extremely valuable when it is time to interpret hypothesis testing results.

Putting it All Together

It is important to remember what has happened in the last two chapters. We have been examining how a distribution of data is measured. Figure 7.5 provides a visual of everything we have learned about how distributions of data are measured. Measures of central tendency (mean and median) tell you the characteristics of the middle of the data. Measures of dispersion (range, sd, quartiles, percentiles) tell you the spread of the data. Skew tells you if the data leans to the left or right and kurtosis tells you the peakness of the data. Every statistics seen so far has provided insight into what the data looks like.

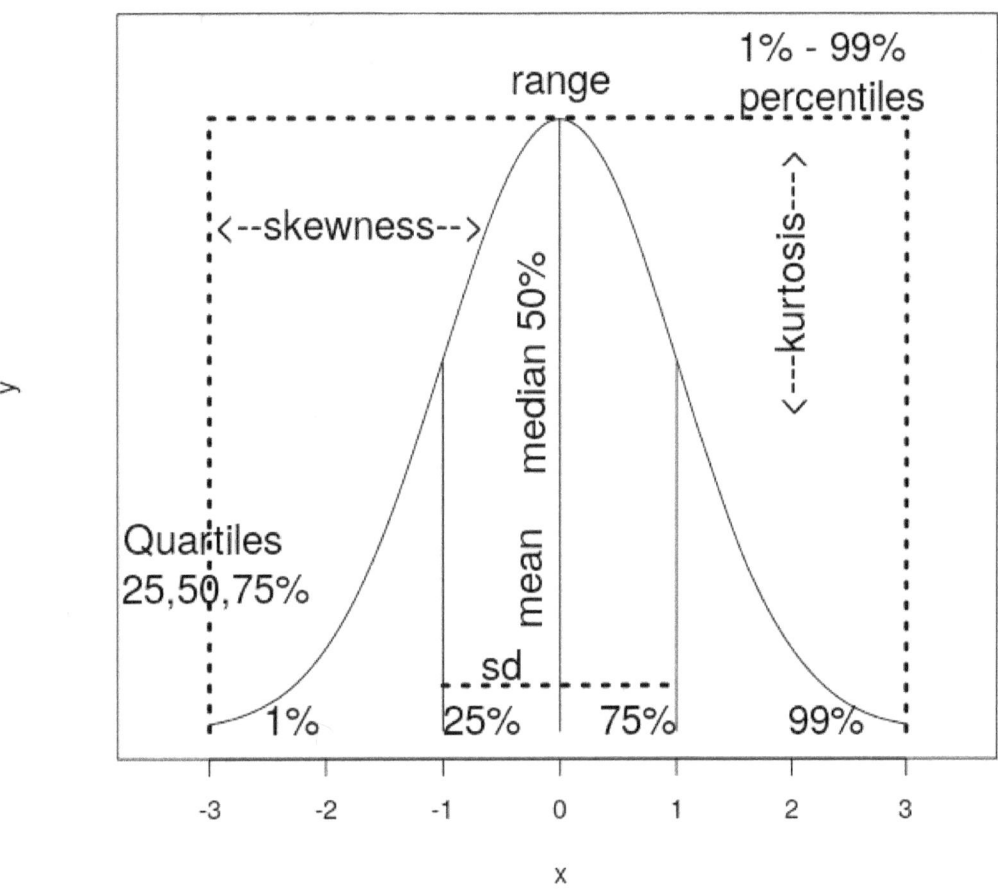

Figure 7.5: Characteristics of a Distribution
These measures of central tendency and dispersion are similar to how we describe people. Some people are tall or short and we can think of this as kurtosis. Some people are left-handed or right handed and we can think of this as skewness. Some people are really smart or dumb and we can think of this as dealing with percentiles and quartiles. To simplify this just think of these statistical tools as ways to describe data just as adjectives are used to describe people

Conclusion

The normal distribution is a tool that allows us to determine the probability of an event. This is based on an assumption that the data is normally distributed. This is why we check skew and kurtosis so that we can determine if the data is normally distributed. When we know the normality of our data, we do not really use the normal distribution but instead use the standard normal distribution. This allows for the comparison of different data with different scales by rescaling all data to a mean of 0 and a standard deviation of 1. When employing standard normal distribution we are finally able to determine probabilities by comparing the z-score to the distribution table.

Points to Remember

- The normal distribution is the most commonly used distribution in statistics

- The standard normal distribution sets the standard or mean to 0 and the deviation or dispersion to 1.

- Standard normal distribution is used for determining z-scores which in turn can be used for determining probabilities

R Code Used

- pnorm(): Converts z-score to probability.

Exercises

1. Find the area under the standard normal distribution curve between z = 0 & z = 1.45.

2. Find the area under the standard normal distribution curve between z = 0 & z = 0.45.

3. Find the area under the standard normal distribution curve between z = 0 & z = -1.20.

4. Find the area under the standard normal distribution curve between z = 0 & z = -2.70.

5. Find the area under the curve to right of z = 1.67

6. Find the area under the curve to right of z = 1.06

7. In the cars dataset, find the probability that the speed of a car is greater than 19.

8. In the cars dataset, find the probability that the weight of a car is greater than 70

Chapter 8

What are Confidence Intervals?

In this chapter, you will learn the following...

- The purpose of confidence intervals
- How to calculate confidence intervals

Key Terms.
Confidence interval   Alpha   Max error of the estimate

Introduction

In this chapter, we look at confidence intervals. Confidence intervals are used extensively in inferential statistics to provide insights into what to expect when trying to understand parameters at the population level. Almost all statistics can have a confidence interval calculated for them.

Understanding Confidence Intervals

You can see that confidence interval involves two words, which are the word "confidence" and "interval." First, we will look at what the word interval means in statistics. In statistics intervals is an estimate that includes a range of values. Rather than picking a specific point, an interval indicates that the value you are looking for is somewhere between two points. For example, if you find that the confidence interval for the mean of the average height of male university students is 170-178cm. What you are saying is that the specific average height of a male university student is somewhere between 170-178cm. This is what is meant by the interval part of the confidence interval.

For the word "confidence", this word indicates how sure or confident we are of our interval estimate. In statistics, we are normally 90%, 95%, or 99% confident in our interval estimate as these are the traditional cutoffs in research. All this has to do with probability or likelihood of something happening. By stating 95% confidence it is saying that if we were to repeat the measurement of the height of university male students about 95% of the time we would get means within the confidence interval. In other words, we are beginning to infer about the population through what we can find from the sample. This is the beginning of inferential statistics.

The higher the confidence the wider the interval will be. For example, returning to our male university students heights, if we are 95% confident the interval might be 170-178cm. However, if we are 99% confident it is necessary to have included a wider interval to accommodate the higher confidence. Higher confidence means a wider interval in order to achieve the certainty that you are looking for.

This use of 90%, 95%, and 99% is known as the confidence level. The calculation of the confidence level is as follows.

$$(1 - a)100$$

a = significance level

'a' is the alpha or significance level, which is an indication that what is being observe is not random chance. If you look closely at the numbers, you may be able to see that we normally set alpha value at .1, .05, .01. When alpha is set to .1 it is determining a 90% confidence interval, .05 is determining a 95% confidence interval and .01 is determining a 99% confidence interval.

We normally, divide alpha by 2. The reason for this is that we need to accommodate both sides or tails of our normal distribution. In a distribution there is always to tails. These parts are near the edges of the distribution. Below in Figure 8.1 you can see what is meant by tails.

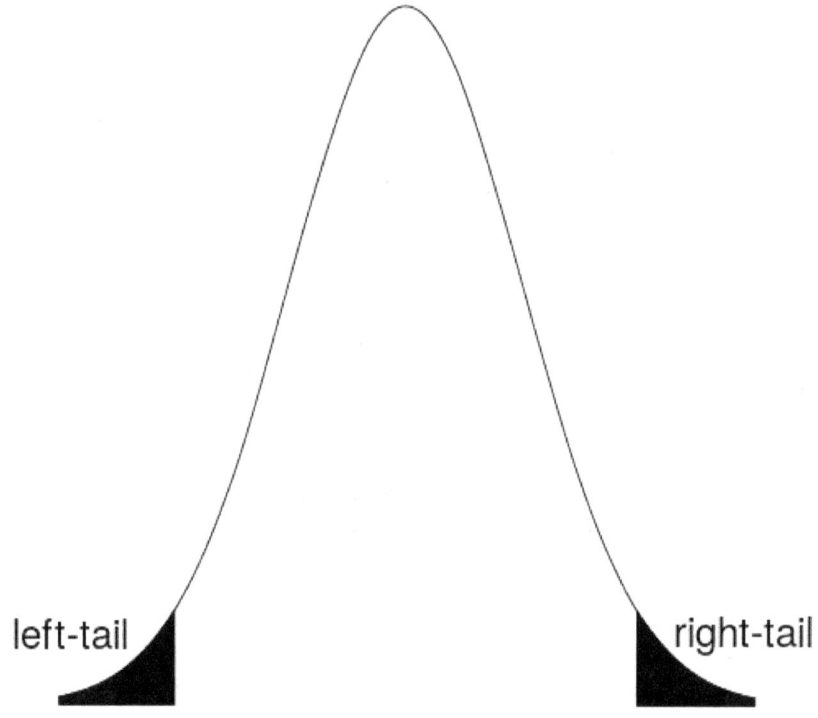

Figure 8.1: Tails of a Distribution

If we set a = .10 we need to divide .10 by 2 to get .05 for both tails of the distribution. For an alpha of .05 both sides of the tail get .025 and for .01 both tails get .005.

I want to focus specifically on the 90% confidence interval which is set with an alpha of .1. To figure out the z score, which is what we use for the standard normal distribution, we subtract .05 from .5 to get a 0.45. We subtract from .05 because we divide the original alpha by 2 so we must also divide the distribution by 2 or 1/2 = .5. The number 0.45 is then converted to the z-score by looking at the z-score table which gives us 1.65. Below in figure 8.2 and 8.3 are the general instructions of the steps we took.

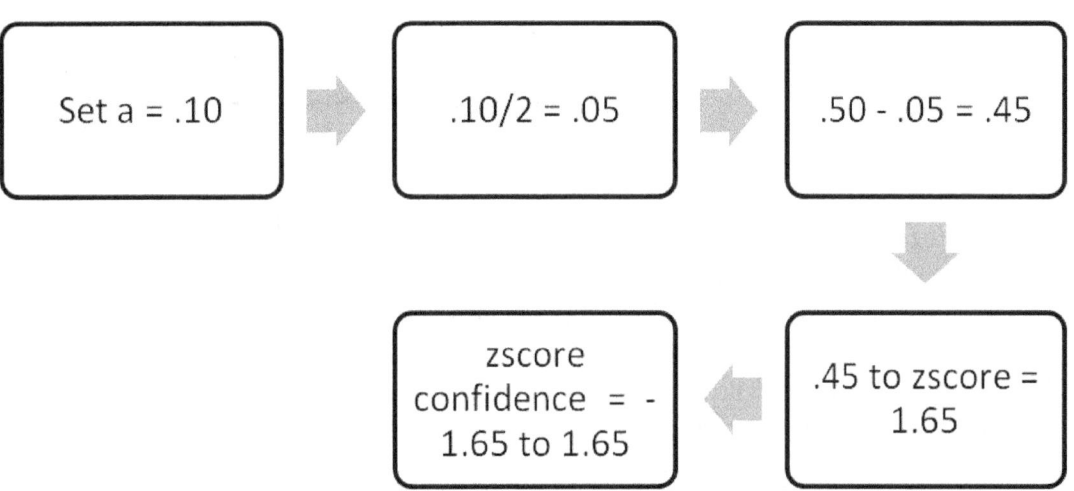

Figure 8.2: Specific Example of Calculating Confidence Interval

CHAPTER 8. WHAT ARE CONFIDENCE INTERVALS?

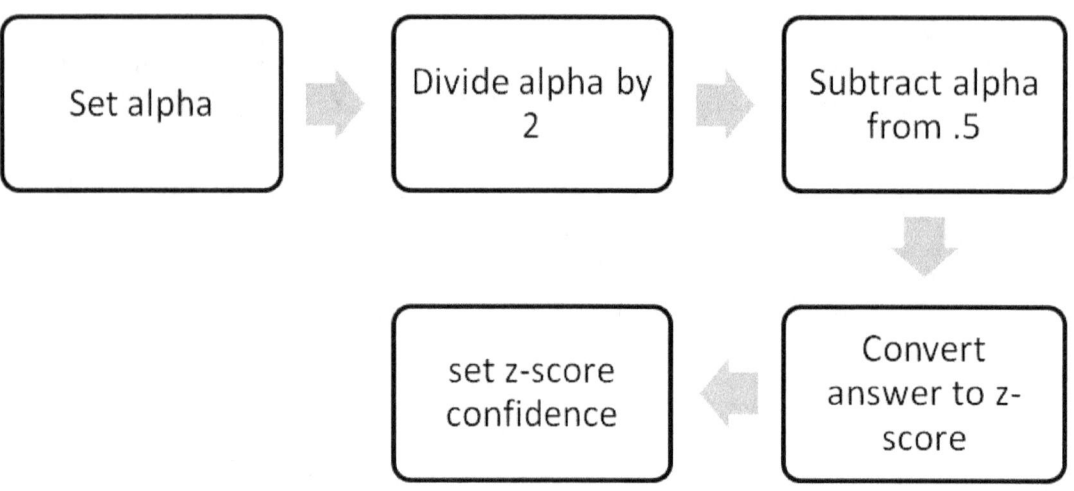

Figure 8.3: Steps to Find Confidence Interval

Another we to think about confidence intervals is that the 95% confidence interval is a range that is 2 standard deviations above and below the mean. In addition, the 99% confidence interval has a range that is 3 standard deviations above and below the mean. In other words, we have already looked at confidence intervals without knowing it. Table 8.1 shows the confidence intervals for the Sepal.Length variable in the iris dataset.

This chapter is simply providing the details on how to calculate them. However, it is rare to have to calculate confidence intervals manually. The purpose here is to understand what the computer is doing for you. Figure 8.4 is a diagram of a 90% confidence interval.

Standard Deviation	Below mean	Above mean	Range
1 (68%)	5.84 - .82 = 5.02	5.84 + .82 = 6.66	5.02 - 6.66
2 (95%)	5.84 - $(.82^2)$ = 4.20	5.84 + $(.82^2)$ = 7.48	4.20 - 7.48
3 (99%)	5.84 - $(.82^3)$ = 3.38	5.84 + $(.82^3)$ = 8.30	3.38 - 8.30

Table 8.1: Confidence Interval Table

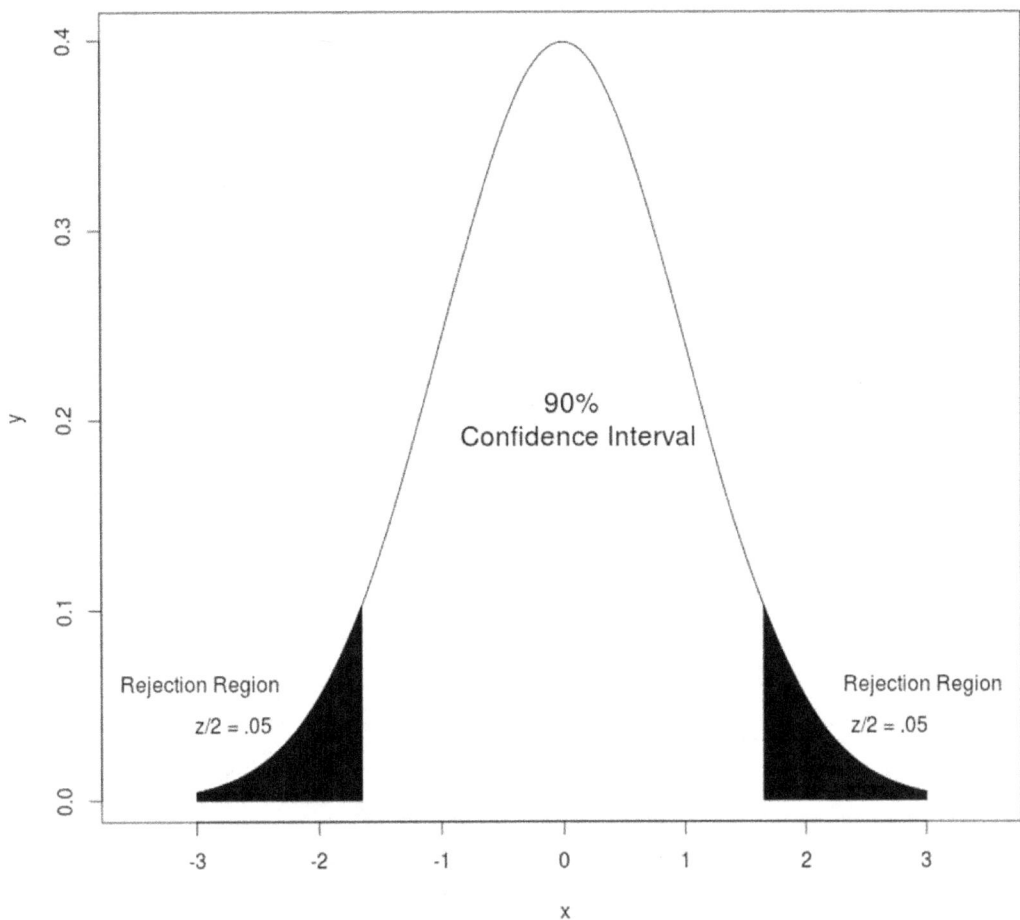

Figure 8.4: 90% Confidence Interval
Our point is somewhere in the white region above. We must have an upper and lower limit in order to have a confidence interval. If we placed the entire black region on one side, our point could be anything except the black region, which could reach into infinity as shown below in Figure 8.5.

CHAPTER 8. WHAT ARE CONFIDENCE INTERVALS?

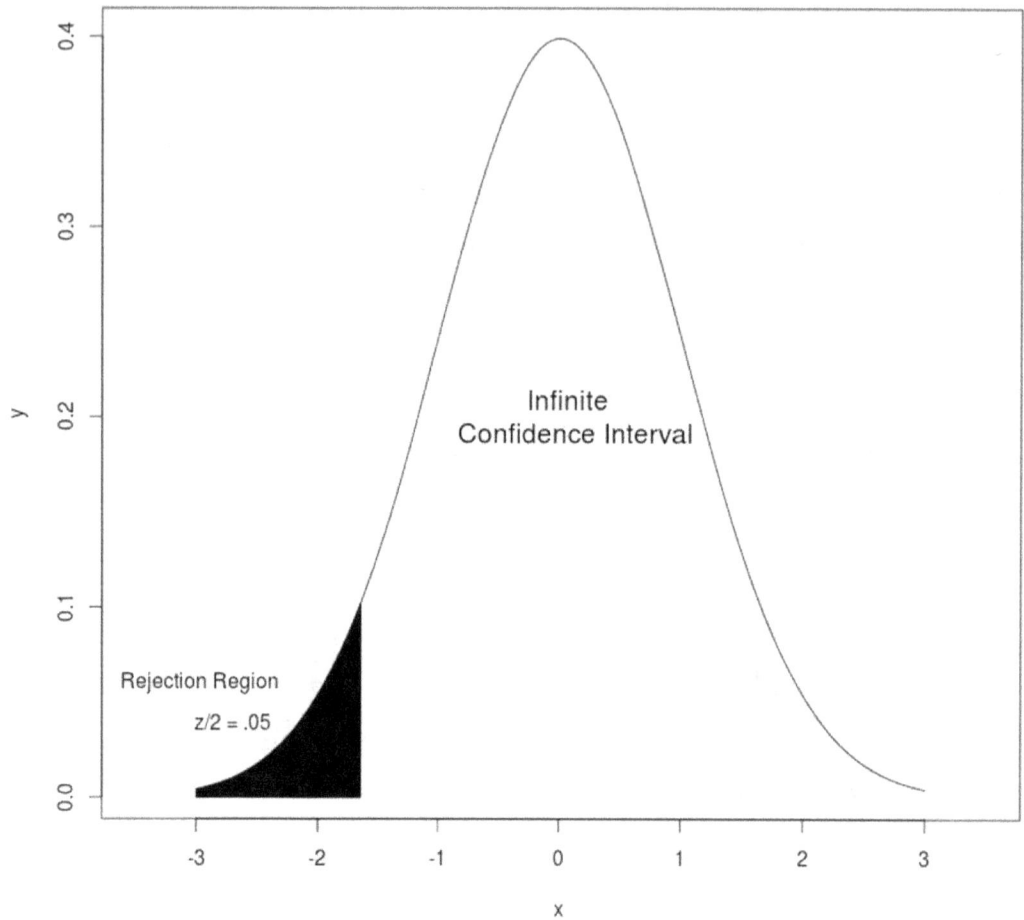

Figure 8.5: Infinite Confidence Interval (In White Region)

The lack of an upper limit means are point can be anything from -1.65 to infinite. This leads to no confidence. This is also why we divide alpha by 2. It allows us to set limits to our confidence. Table 8.2 contains the three commonly used confidence intervals with the alphas and the z-scores.

1-a	a	a/2	Za/2
.90	.10	.05	1.65
.95	.05	.025	1.96
.99	.01	.005	2.58

Table 8.2: Alpha and Z-Scores for Confidence Intervals

Below is the actually equation for finding a confidence interval.

$$\bar{X} - z_{a/2}\left(\frac{s}{\sqrt{n}}\right) < \mu < \bar{X} + z_{a/2}\left(\frac{s}{\sqrt{n}}\right)$$

This equation is known as the maximum error of the estimate. Using this equation, you need to know the sample mean, sample size, and standard deviation to find the confidence interval of the mean. Below is an example.

Example 8.1

55 students took a quiz in stat class. The average score was 73 with a standard deviation of 7.3. Find the 95% confidence interval.

$$73 - 1.96\left(\frac{7.3}{\sqrt{55}}\right) < \mu < 73 + 1.96\left(\frac{7.3}{\sqrt{55}}\right)$$
$$73 - 1.93 < \mu < 73 + 1.93$$
$$71.07 < \mu < 74.93$$

We are 95% confident that the population mean is between 71.07 and 74.93. In simple terms, if we repeated the sample process 95% of the samples confidence intervals would contain the population mean.

Finding Confidence Intervals Using R

For R, we will find the 95% confidence interval for the mean of the Sepal.Length variable in the iris dataset. You will need to install the Rmisc package and the CI function. The code is below.

```
> CI(iris$Sepal.Length,ci=.95)
   upper     mean     lower
5.976934 5.843333 5.709732
```

The results indicate that we are 95% confident that the mean of the Sepal.Length variable is between 5.7 and 6. You use the CI function by placing the name of the variable inside the parentheses followed by the desired confidence interval.

CHAPTER 8. WHAT ARE CONFIDENCE INTERVALS?

Confidence Interval for Proportions-Categorical

Confidence intervals can also be calculated for proportions, which are used for categorical data. A proportion is a part of a whole, which is different form of measurement from our previous examples. Proportions are always represented as a decimal or percentage calculated from the total number of observations. You calculate the proportion as shown below.

$$\hat{p} = \frac{x}{n} \text{ and } \hat{q} = 1 - \hat{p}$$

$\hat{p}$ = Sample proportion
$\hat{q}$ = Remaining proportion
x = Observations
n = sample size

The actual equation for finding the confidence interval is below.

$$\hat{p} - z_{a/2}\sqrt{\frac{\hat{p}\hat{q}}{n}} < p < \hat{p} + z_{a/2}\sqrt{\frac{\hat{p}\hat{q}}{n}}$$

p = proportion

Below is an example using proportion confidence intervals.

Example 8.2

A sample of 235 students applied to study at a university including 123 women. Find the 95% confidence of the proportion of women who applied to university. Here is what we know
X = 123
n = 235
a = 1.96
$\hat{p} = \frac{123}{235} = .52$
$\hat{q} = 1-.52 = .48$

$$.52 - (1.96)\sqrt{\frac{(.52*.48)}{235}} < p < .52 + (1.96)\sqrt{\frac{(.52*.48)}{235}}$$
$$.52 - 0.06 < p < .52 + 0.06$$
$$.46 < p < .58$$

Our answer is that we can be 95% confident that the proportion of female applicants is between 0.46 and 0.58.

Finding Proportion Confidence Intervals Using R

To find the confidence interval of a proportion in R you need to first install the PropCIs package. You use the blakerci function. Then you simple plug in the values for x, n, and set the confidence interval as shown below. We are using the same example data.

```
> blakerci(x = 123,n = 235,.95)
data:
95 percent confidence interval:
 0.4594143 0.5875765
```

The results above are the same as what we calculated manually.

Conclusion

Confidence intervals provide you with an idea of the typical dispersion of the data. This is useful information for making inferences about results. Any sample is going to have error and other issues. The power of confidence intervals is that they help to strengthen whatever conclusion you are considering as you make decisions based on data.

Points to Remember

- Confidence intervals are a measure of the dispersion around a statistic
- Confidence intervals are simply a calculation of the distance by standard deviations from a statistic
- Confidence intervals can be calculated for continuous or categorical data.

R Code Used

- CI(): Calculates confidence interval.
- blakerci(): Calculates confidence interval of proportions.

Exercises

1. A total of 35 students took a intro to statistics. The average grade was 77 with a standard deviation of 12.7. Find the 95% confidence interval.

2. A total of 20 students had their weight taken. The average weight was 80kg with a standard deviation of 4.7. Find the 99% confidence interval.

3. A total of 85 students took basic ESL entrance exam. The average score was 5.7 with a standard deviation of 0.7. Find the 90% confidence interval.

4. Use R to find the 95% confidence interval of the speed variable in the cars dataset

5. Use R to find the 95% confidence interval of the weight variable in the cars dataset

6. A sample of 400 students participated in a nationwide elections with 254 voting for the new president. Find the 95% confidence of the proportion of people who voted for the president

7. A sample of 150 students found that 128 passed stats. Find the 90% confidence of the proportion of students who passed stats

8. Use R, a sample of 200 students found that 73 wear motorbike helmets. Find the 95% confidence of the proportion of people who wear motorbike helmets

9. Use R, a sample of 130 found that 45 are overweight. Find the 90% confidence of the proportion of people who are overweight.

Chapter 9

What is a Hypothesis Testing?

In this chapter, you will learn the following...

- The purpose hypothesis testing
- How to calculate one-sample t-test
- How to t-test for proportion

Key Terms.
  Hypothesis test   Null Hypothesis   Alternative Hypothesis

Introduction

There are times in statistics when we want to compare two values and determine the probability that they are different or if the difference is just by chance. This is called hypothesis testing and is a key cornerstone of statistics. To appreciate hypothesis testing you must have a basic understanding of probability, standard normal distribution, z-scores, and significance level.

The values you want to test have to be of the same statistic. In other words, you compare means with means, skew with skew, etc. In addition, you need to know if you are comparing your statistic to another sample statistic or to a distribution such as the normal distribution. In this chapter, we will compare a sample mean to what would be expected in the distribution of the population.

In order to conduct a hypothesis test you first must decide what your hypotheses are. There are always two hypotheses in hypothesis testing and these are the null (H0) and alternative hypothesis (H1).

- The null hypothesis states that there is **no difference** in the two values

- The alternative hypothesis stats that there is **a difference** between the two values

In statistical research there are no other choices when it comes to hypotheses. It is all no difference or difference.
Below is an example situation followed by the null and alternative hypothesis.

Example 9.1

You find that the average weight of students is 65kg. The expected value was 76kg. What is the null and alternative hypothesis for this situation?

- Null: There is no difference between the students weight and the expected weight

- Alternative: There is a difference in between the students weight and the expected weight

It is important to note that you only test the null hypothesis in statistics. In other words, you are always testing under the assumption that there is no difference between your two values. Furthermore, you only reject or do not reject a null hypothesis you never accept anything as you are never 100% sure of your results.

Next, we need to pick the significance level or alpha. Before we do this we need to review some information that we have learned in order to see the connection between the various chapters of the book. We began this book by learning about the mean and other values of central tendency. The mean is a crucial value in statistics that serves as a foundation for calculating many other values such as standard deviation, which is a value that measures dispersion.

The standard in standard deviation is the mean. The deviation is how much on average an individual data is different from the standard aka the mean. We use the mean to find the standard deviation. With our knowledge of a value of central tendency (mean) and a value of dispersion (standard deviation) we can use this for calculating the probability of a range of values in a continuous distribution.

The standard deviation is used in particular for the standard normal distribution. This is the bell-shaped curve that is commonly used in statistics. Remember that values towards the middle are more common than values towards the edge. In addition, you can divide up the standard normal curve so that 68% of the values are within one standard deviation of the mean 95% are within two standard deviations and 99% are within three standard deviations. Percent here can also mean probability.

In the previous chapter, we calculated various ranges of probability in relation to confidence intervals. Our goal was to determine the range of values in the unshaded region. If you remember you know that we can calculate a 90%, 95%, and 99% confidence interval. This means that 10%, 5%, and 1% of the distribution was darkened.

Hypothesis testing combines our knowledge of mean, standard deviation, and confidence intervals into one practical application. With hypothesis testing, the goal is to determine if a value falls in the unshaded region called the non-rejection area or in an area called the rejection region or the darkened region when calculating confidence intervals. Figure 9.1 provides a visual of the knowledge that leads to hypothesis testing.

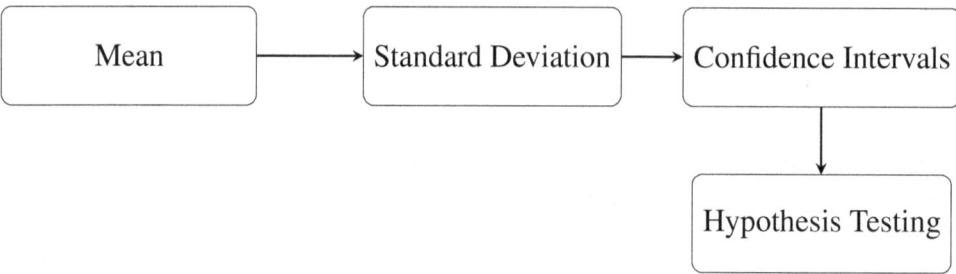

Figure 9.1: Background to Hypothesis Testing

For hypothesis testing, the rejection region is all on one side or the other or split equally on both sides of a standard normal distribution. Generally, we have the choice of a one-tail or two-tail test. A one-tail test places the rejection all to the left (below the mean) or to the right (above the mean) in the distribution. The two-tailed test divides the rejection zone equally on both sides. The alpha allows us to setup the rejection region in our standard normal distribution. Generally, the alpha is set to 0.01, 0.05, or 0.1. If we take 0.05 as an example we are saying that there is a 5% or less chance that the results happened by random. Since this is such a low probability, it is one reason that we would reject the null hypothesis.

If the probability value is greater than 0.05 or 5%, we would not reject the null hypothesis because the difference we may see may be due to chance or luck. The alpha is a minimum cutoff point while the probability value is the actual number that R returns to you when you do the analysis. The shorten name for probability value is pvalue.

Let's go through an example, let's say that you compare the heights of men and women at a university and you get a pvalue of 0.03 or 3%. What this literally means is the that there is a 3% chance or probability that there is no difference in the heights of men and women. Since this is a small chance and less than 0.05 or 5% we reject the null hypothesis that there is no difference in the height of men and women.

What this means in simply terms is that since we reject the idea there is no difference in height between men and women we need to figure out how to explain this. The simply answer is to say that gender makes a difference in the heights of man. This conclusion is based on the variables we measured. We measured height and we also determine the genders of the people in the study. Gender effects height or height effects gender.

The alpha that is chosen determines the size of your rejection region. The bigger the alpha the bigger the rejection region. As such, an alpha of 0.1 is ten times bigger than a alpha of 0.01. When you do your analysis, R will automatically convert the z-score to the pvalue.

Something else to mention is that the null hypothesis can be stated three different ways.

- There is no difference between the sample mean and expected mean (two-tails)
- The mean of sample mean is greater than the expected mean (one-tail)
- The mean of sample mean is less than the mean expected mean(one-tail)

Figure 9.2 is what one-tailed rejection regions looks like when the p-value is set to 0.05. In a one-tail test, if the rejection region is to the left, you are saying the following in your hypothesis if we reuse Example 9.1.

- Null: there is no difference between the means
- Alternative: The student mean is less than the expected mean

CHAPTER 9. WHAT IS A HYPOTHESIS TESTING?

Any score in the black range would indicate we should reject the null hypothesis. The starting point of the black region is called the critical value as mentioned earlier.

If the rejection region is to the right here is what we are saying now.

- Null: there is no difference between the means

- Alternative: The expected mean is greater than the students' mean

Whether you go with the rejection region to the left or right depends on your perspective. You need to decide what you are comparing and what is considered normal in that situation.

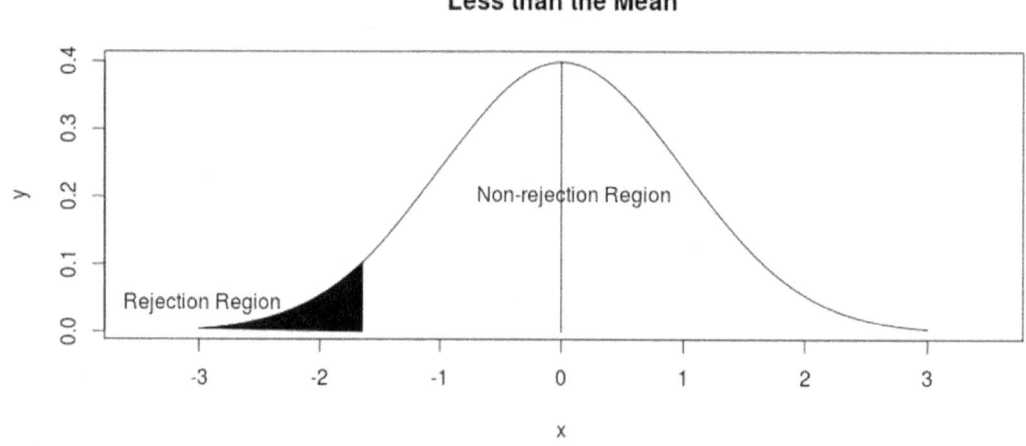

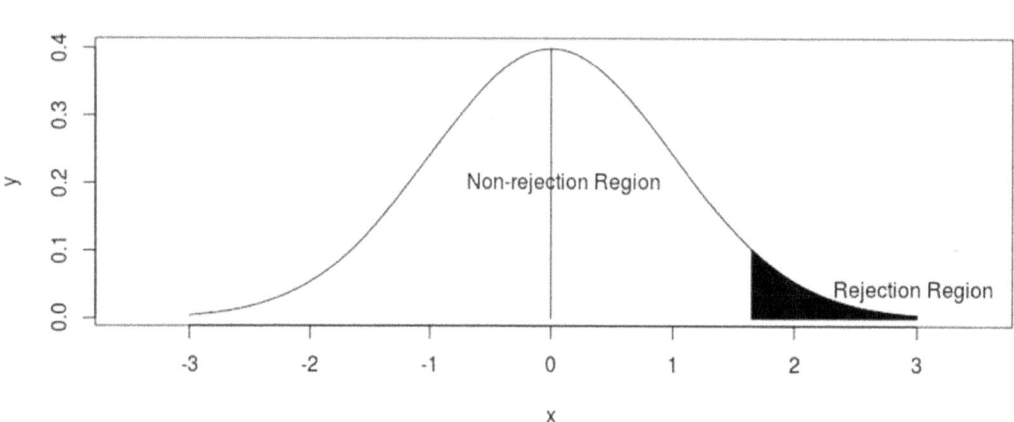

Figure 9.2: One-Tail Rejection Regions

For a two-tail test, you place half the critical value on one side and half on the other side. This is because now it is possible for the mean to above or below. Since there are two possibilities, we need two rejection regions. Therefore, instead of one region that is 5% we have two regions that are 2.5% each. Figure 9.3 shows what this looks like.

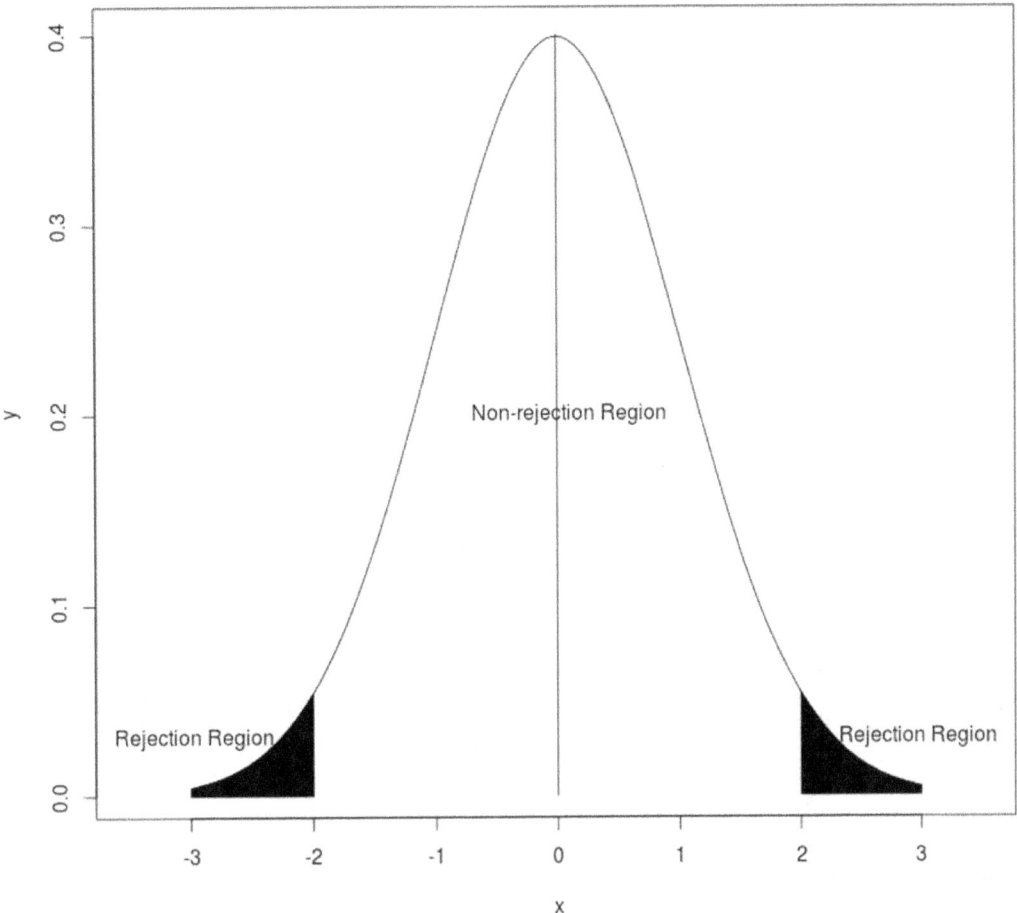

Figure 9.3: Two-Tail Rejection Regions

What this means is that we will reject any value that is in the top 2.5% or the bottom 2.5% as being different from what we would expect from a normal distribution

In terms of hypotheses, this is what we are saying here if we reuse Example 9.1 is the following

- There is no difference in the means
- There is a difference between the means

Table 9.1 critical values for both a one-tail and two-tail test.

Significance	One-Tail	One-Tail Critical Value	Two-Tail	Two-Tail Critical Value
.90	.10	1.28	.05	1.65
.95	.05	1.65	0.025	1.96
.99	.01	2.32	0.005	2.58

Table 9.1: One & Two-Tail Critical Values

We will now turn our attention to a concrete example starting with the one sample t-test.

One Sample t-test

When you are using the standard normal distribution you are also calculating z-scores to find probabilities. However, most of the time in statistics, we use the t-distribution. The reason for this is that the t-distribution is useful for small sample sizes below 30. However, if your sample size is above 30 the t-distribution is almost exactly the same shape as the standard normal distribution. Therefore, the t-distribution is used most commonly for calculating probabilities. The main difference is that the t-distribution's shape changes slightly based on the sample size. If you want to convert the t-stat to a probability manually you can look at the appendix.

Our first statistical test is a one sample t-test. This test compares the mean you found in your sample and compares it to the expected value from the population. One sample means that you have only one sample and you are comparing a value from that one sample to the population. This test is used when you know the following

- The sample mean
- The mean of the population
- Standard deviation of the mean
- Sample size

Here are the steps for conducting a one-sample t-test

1. State the hypotheses
2. Select the alpha
3. Determine the critical value
4. Calculate the one sample t-test value (This is found looking a t-distribution tables. These are available online and R calculates this value for you automatically)

5. Make decision

6. Draw conclusion

Let's begin with a one-tail test

Example 9.2

A report finds that the average monthly salary for a daily worker in Thailand is more than $5,110 per year with a standard deviation of 234 dollars. A sample of 47 workers has a mean of $5,200 per year. With a = 0.05, determine if the sample of workers earn more than the population mean.

Step 1: State the hypotheses
H0 = There is no statistical difference between the population mean of 5,110 and the sample mean of 5,200
H1 = The sample mean of 5,200 is statistical different from the population mean of 5,110

Step 2: Select the alpha
a = 0.05

Step 3: Determine the Critical Value
for a one-tail test, the t-critical value is 2.32

Step 4: Compute Compute the one sample t-test

$$t = \frac{\bar{x} - \bar{\mu}}{\sigma/\sqrt{n}} = \frac{5200 - 5110}{235/\sqrt{47}} = \frac{90}{34.27} = 2.62$$

Step 5: Decision We reject the null hypothesis at a = 0.05, since 2.62 > 2.32

Step 6: Conclusion
Since we reject the null hypothesis we can conclude that the average salary of the daily workers is different from the population mean.

CHAPTER 9. WHAT IS A HYPOTHESIS TESTING?

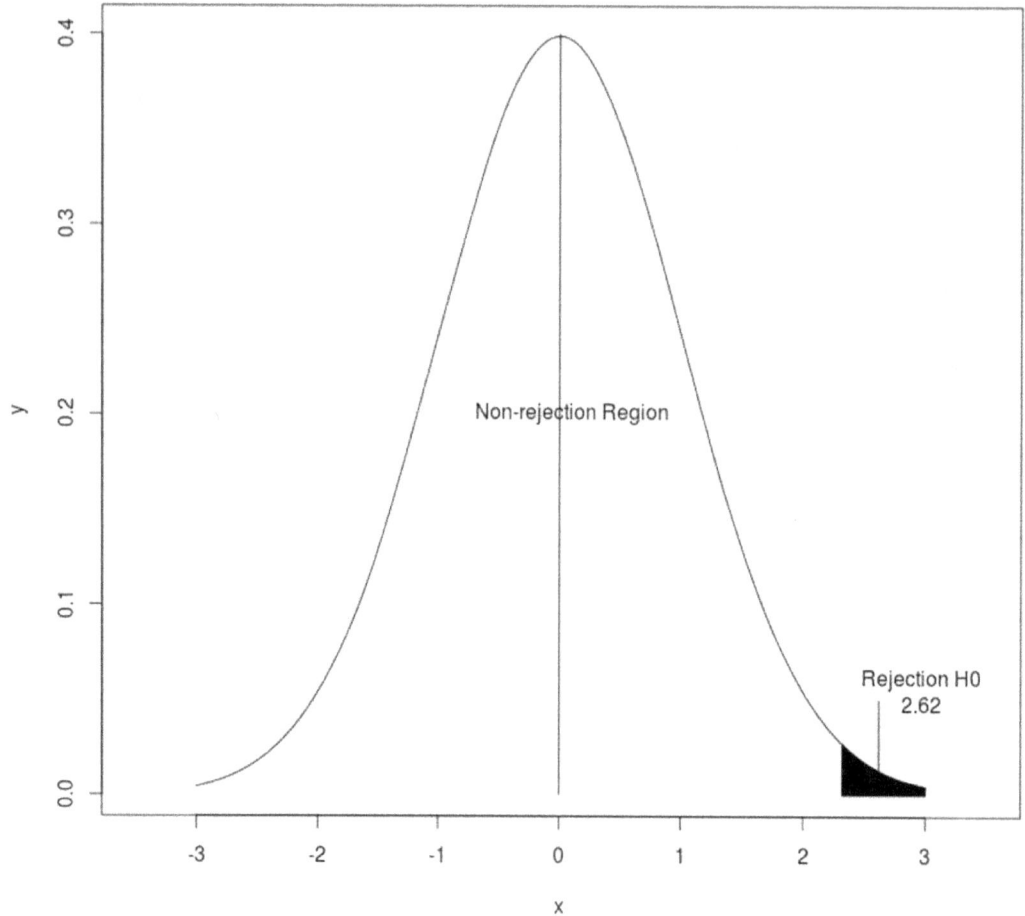

Example 9.2

Example 9.3

The average class has 13 students and standard deviation of 3. A sample of 25 classes had a mean of 12. Determine if the mean of the sample is different from the population at a = 0.05.

Step 1: State the hypotheses
H0 = There is no statistical difference between the population mean of 13 and the sample mean of 12
H1 = The sample mean of 12 is statistical different from the population mean of 13

Step 2: a = 0.05

Step 3: For a two-tail test, the z critical value is ±1.96

Step 4: Compute the one sample z test

$$t = \frac{\bar{x}-\bar{\mu}}{\sigma/\sqrt{n}} = \frac{12-13}{3/\sqrt{25}} = -\frac{1}{.6} = -1.66$$

Step 5: We do not reject the null hypothesis at a = 0.05, since -1.66 > -1.96

Step 6: Since we do not reject the null, we can conclude that the average class size of the population 12 and the mean of the sample 13 students per class are the same.

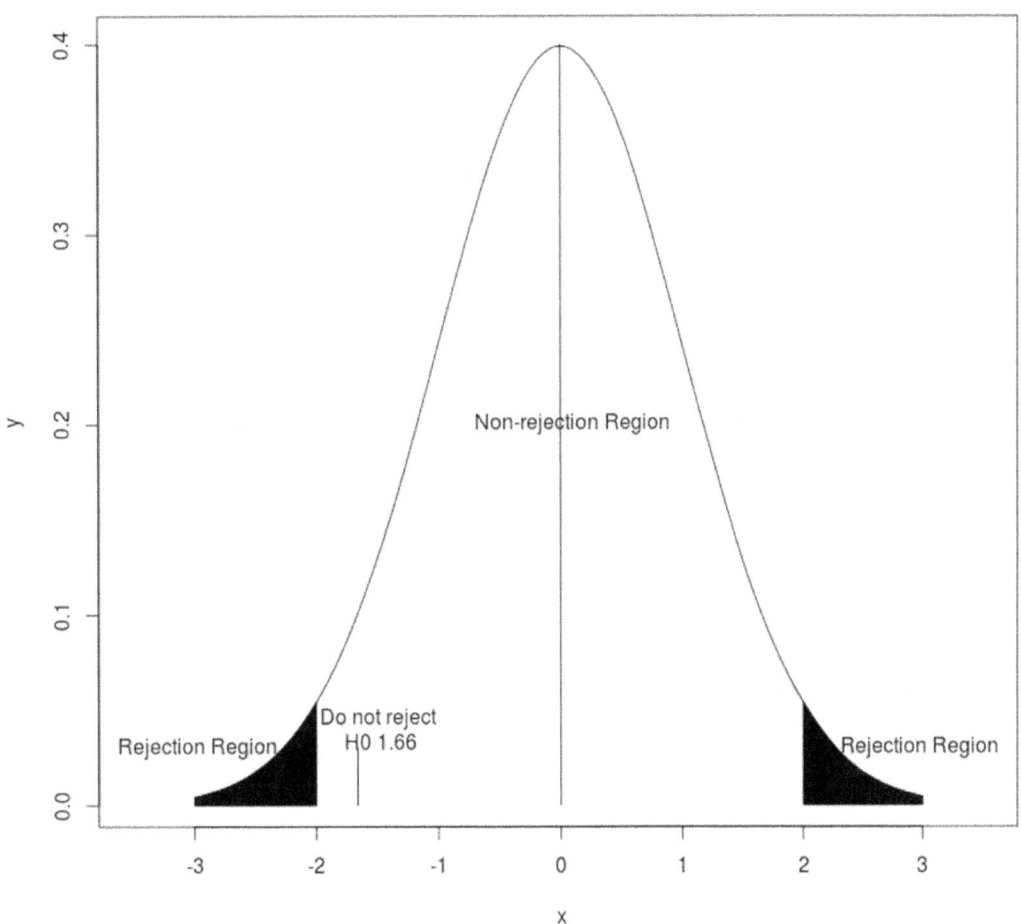

CHAPTER 9. WHAT IS A HYPOTHESIS TESTING?

One Sample t-test Using R

Below is how we do this in R. When calculating the one sample t-test in R you must have the raw data of the sample. What this means is that you cannot just type in the mean but you must rather input the actually data. The code below should make this clearer.

```
> classMean<-
c(10,10,10,10,10,10,10,10,10,10,10,16,16,16,16,16,16,16,16,16,16,13,14
,14,14)
> t.test(x=classMean,mu=13,sd=3,n=25, alternative = 'two.sided')

        One Sample t-test

data:  classMean
t = 0, df = 24, p-value = 1
alternative hypothesis: true mean is not equal to 13
95 percent confidence interval:
 11.83248 14.16752
sample estimates:
mean of x
       13
```

The answer are not identical because we had to use the raw data here. However, the R output includes the pvalue as well as the confidence interval. Below are the results for the one-sided test in example 9.2.

```
> workerMean<-
c(5000,5000,5000,5000,5000,5000,5000,5000,5000,5000,5000,5000,5000,5000,
+       5000,5000,5000,5000,5000,5000,5000,5000,5000,5000,5400,5400,5400,5400,
+       5400,5400,5400,5400,5400,5400,5400,5400,5400,5400,5400,5400,5400,5400,
+       5400,5400,5400,5400,5600)
> t.test(x=workerMean,mu=5110,sd=234,n=47, alternative = 'greater')

        One Sample t-test

data: workerMean
t = 2.9591, df = 46, p-value = 0.002431
alternative hypothesis: true mean is greater than 5110
95 percent confidence interval:
 5148.944    Inf
sample estimates:
mean of x
    5200
```

We use the t.test function along with several arguments to calculate the value. Notice how the confidence interval ends in infinite (inf) this is because this was a one-sided test. When the test is one-sided the confidence interval cannot really be calculated.

t-test for proportion: Categorical Data

It is also possible to calculate t-stat for comparing proportions. We have covered proportions in a prior chapter. The steps to take are mostly the same with the use of percentages and no standard deviation. Below is the equation

$$t = \frac{\hat{p}-p}{\sqrt{pq/n}}$$

Example 9.4

A survey found that 78% of the students pass grammar class. A sample of 80 students found that 76% passed grammar class. At alpha = 0.05 is the sample of 76% less than the population of 78%

Step 1: Set hypothesis
H0 = There is no statistical difference between the population of proportion of 78% and the sample proportion of 76
H1 = The sample mean of 76% is statistical less than population proportion of 78%

Step 2: Set alpha a = 0.05

Step 3: Set Critical Value -2.32

Step 4: Calculate t stat

$$t = \frac{\hat{p}-p}{\sqrt{pq/n}} = \frac{76-78}{\sqrt{(.78*.22)/80}} = -\frac{2}{1.17} = -1.71$$

Step 5: Decision
We do not reject the null hypothesis at a = 0.05, since -1.71 > -2.32

Step 6: Conclusion
Since we do not reject the null hypothesis, we conclude that the sample mean is not less than the population mean.

CHAPTER 9. WHAT IS A HYPOTHESIS TESTING?

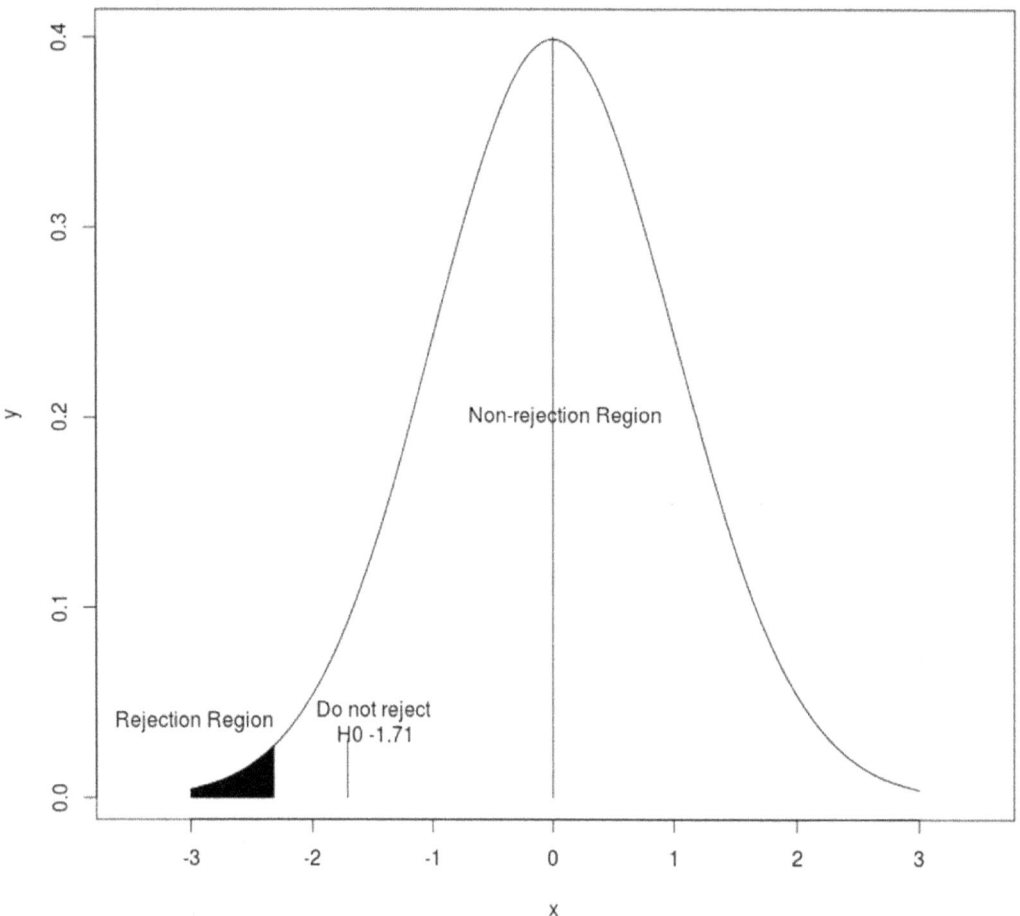

Example 9.4

Example 9.5

A survey found that 27% of the students skip class. A sample of 50 students found that 38% skip. At alpha = 0.05 is the sample of 38% different from the population of 27%

Step 1: Set hypothesis
H0 = There is no statistical difference between the population of proportion of 27% and the sample proportion of 38%
H1 = The sample proportion of 38% is statistical different from the population proportion of 27%

Step 2: Set Alpha
a = 0.05

Step 3: Set Critical Value
Critical value is ±1.95

Step 4: Calculate z score

$$t = \frac{\hat{p}-p}{\sqrt{pq/n}} = \frac{38-33}{\sqrt{(.27*.73)/65}} = \frac{5}{3.5} = 1.43$$

Step 5: Decision
We do not reject the null hypothesis at a = 0.05, since 1.43< 1.95
Step 6: Conclusion
Since we do not reject the null hypothesis, we conclude that the sample mean is not less than the population mean.

Below is the R code for the last two examples

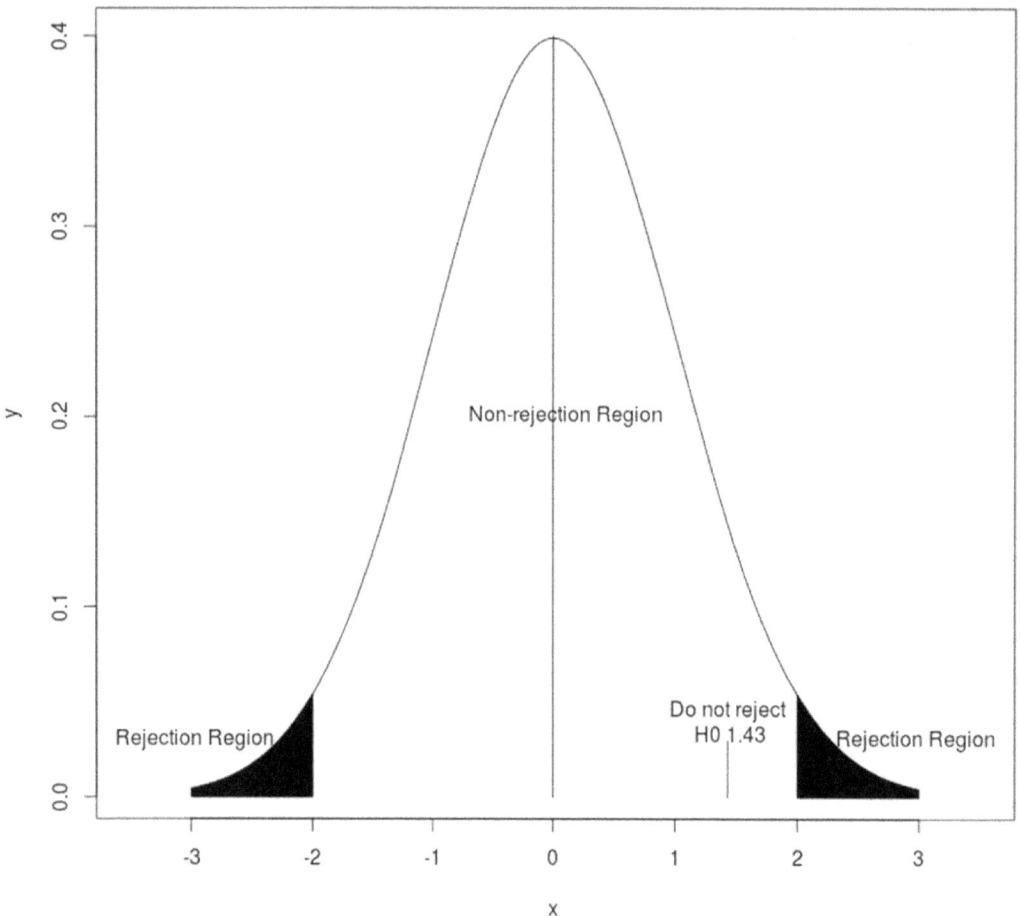

Example 9.5

Test of Proportions Using R

Example 9.4

```
> prop.test(x=61,n=80,p=0.78,alternative = 'less')

    1-sample proportions test with continuity correction

data:  61 out of 80, null probability 0.78
X-squared = 0.059003, df = 1, p-value = 0.404
alternative hypothesis: true p is less than 0.78
95 percent confidence interval:
 0.0000000 0.8367229
sample estimates:
     p
0.7625
```

Example 9.5
```
> prop.test(x=19,n=50,p=0.27,alternative = 'two.sided')

    1-sample proportions test with continuity correction

data:  19 out of 50, null probability 0.27
X-squared = 2.5368, df = 1, p-value = 0.1112
alternative hypothesis: true p is not equal to 0.27
95 percent confidence interval:
 0.2499804 0.5283672
sample estimates:
   p
0.38
```
The coding is slightly different here and is explained below
x = number of success. For us we find this number by finding proportion * sample size.
n = sample size
p = population proportion
alternative= this can be set to greater, less, or two.sided

We use the prop.test function along with several arguments to calculate the value.

Conclusion

This chapter really pulls together what we have looked at so far in this book. Hypothesis testing is when the mean, standard deviations, normal distribution, probability, and confidence intervals work together to allow you to infer from your sample. These concepts are extended in the next chapter when we examine two sample hypothesis testing

Points to Remember

- Hypothesis testing is focused on compare two or more values and determining the probability that they are different from each other.

- There are two types of hypotheses which are null and alternative

- A hypothesis test can identify if one value is greater, less or just different.

R Code Used

- t.test(): Determines if there is a difference between two means

- prop.test(): Determines if there is a difference between two proportions

Exercises

For these questions you need to use the z.test function from the TeachingDemos package. This is because there is no raw data available for these questions which makes it impossible to use the t.test function. Remember that once the sample size is above 30 the t distribution and normal distribution are the same.

1. A study finds that the average salary of students who finished college is less than 15,000 baht per month. At your school, the average salary is 16,200 in a sample of 52 students with a standard deviation of 869. At a significance of 0.05 find out if the students earn less than 15,000 per month

2. A study finds that the average cost of tuition to get a 4-year degree was 369,000 baht. Among 45 colleges in Northern Thailand, the average tuition is 325,000 with a standard deviation of 15,210. At a significance of 0.05 find out if there is a difference between the sample and population mean for tuition.

3. Using R, the average temperature in Thailand was found to be 32 Celsius. Among 38 districts, the average temperature was 27 with a standard deviation of

4. At a significance of 0.05, determine if there is a difference between the sample and population mean.

5. Using R, a report finds that the average rainfall in Thailand is less than 50cm. In 41 cities, the rainfall was 56cm with s standard deviation of 6.4. Determine if the average rainfall is less than 50cm.

6. A survey found that 15% of Thais have a bachelor degree. IN a random sample of 70 Thais 20% had a bachelor degree. At a significance of 0.05 is there a difference between the population and sample proportion

7. A survey found that more than 35% of Thais own a car. In a random sample of 60 Thais 39% owned a car. At a significance of 0.05, determine if the population proportion is more than sample proportion

8. Using R, A survey found that 79% live on campus to study. A random sample of 32 colleges found that 71% live on-campus. At a significance of 0.05 is there a difference between the population and sample proportion

Chapter 10

What is Two Sample Hypothesis Testing?

In this chapter, you will learn the following...

- The purpose of two-sample hypothesis testing
- How to calculate various two-sample t-test
- How to t-test for proportion

Key Terms.
Paired t-test

Introduction

In the previous chapter, we looked at hypothesis sampling for one sample. In that situation, you knew the mean for the population and you want to see if the sample mean was consistent with this. However, there are times in research in which you have two or even more sample means that you take from different populations. In such a situation, you do not know the population mean. Yet, you still want to know if the means of the two groups are different from each other statistically. For example, you may want to compare the quiz scores of men and women or freshmen and sophomores. In each of these examples, you have two groups and you want to determine if there is difference in the scores by group. This is the background for the use of hypothesis testing for two samples.

In addition to having two samples, another aspect to consider is whether the samples are independent or dependent. For example, comparing men and women quiz scores is an example of two independent samples. However, if you sample the same group twice and compare the means this is known as dependent samples because you are comparing a person to their previous performance

T-Test for Two Means

The student's t test was developed by G.W. Gosset in the early 20th century. It is used when the population standard deviation is unknown. Normally, this distribution is the default one used in social science research because we often never know the population standard deviation as we want to generalize our results from a sample.

There are actually two different equations that can be used when calculating the t-statistics. These are the t-stat when variance is assumed equal and the t-stat when variance is assumed unequal. Remember that variance is a measure of the dispersion or spread of the data and that variance is square rooted to determine the standard deviation. Therefore, when variance is assumed equal it means we are assuming the standard deviations are the same. If we assume unequal variance we are assuming that the standard deviations are not the same. Remember that variance is the square of

CHAPTER 10. WHAT IS TWO SAMPLE HYPOTHESIS TESTING?

standard deviation. Below is the equation for the t-stat for assumed unequal variance.

$$t = \frac{(\bar{X}_1 - \bar{X}_2) - (\mu_1 - \mu_2)}{\sqrt{\frac{s_1^2}{n_1} + \frac{s_2^2}{n_2}}}$$

The equation for when variances are assumed equal is

$$t = \frac{(\bar{X}_1 - \bar{X}_2) - (\mu_1 - \mu_2)}{\sqrt{\frac{(n_1-1)s_1^2 + (n_2-1)s_2^2}{n_1 + n_2 - 2}} \sqrt{\frac{1}{n_1} + \frac{1}{n_2}}}$$

As you can see, there is a lot of information here and here is what the variables mean.

$\bar{X}_1$ & $\bar{X}_2$ = sample means of each group

s_1^2 & s_1^2 = variances of each group

n_1 & n_2 = sample size of each group

μ_1 & μ_2 = population mean (unknown and set to 0)

$n_1 + n_2 - 2$ = sets the degrees of freedom which is the number of observations that can vary.

The steps for calculating the t-statistics are as follows

1. State the hypotheses

2. Determine the significance level and degrees of freedom

3. Determine the critical value

4. Compute the t-statistic

5. Make a decision

6. Draw a conclusion

Example 10.1: Two-Tailed Test

You want to know if the salaries of government and international university lecturers are different. The table below includes the mean, standard deviation, and sample size of each group. At a = 0.05 determine if the salaries are the same or not.

Government	International
$\bar{X}_1 = 47,500$	$\bar{X}_2 = 50,100$
$s_1 = 1,500$	$s_2 = 900$
$n_1 = 29$	$n_2 = 25$

Step 1: State hypotheses
H0 = There is no difference in salary
H1 = There is a difference in salary

Step 2: Set alpha and degrees of freedom
a = 0.05, degrees of freedom = 29 + 25 - 2 = 52

Step 3: Find the t critical value
+2.01 two-tailed test.

Step 4: Compute value

$$t = \frac{(\bar{X}_1 - \bar{X}_2) - (\mu_1 - \mu_2)}{\sqrt{\frac{(n_1-1)s_1^2 + (n_2-1)s_2^2}{n_1+n_2-2}} \sqrt{\frac{1}{n_1} + \frac{1}{n_2}}}$$

$$\frac{(47500 - 50100) - 0}{\sqrt{\frac{(29-1)1500^2 + (25-1)900^2}{29+25-2}} \sqrt{\frac{1}{29} + \frac{1}{25}}}$$

$$= -7.65$$

Step 5: Make Decision
Since the t-stat is less than -2.01 we reject the null hypothesis

Step 6: Make Conclusion
There is evidence to consider that there is a difference in salary between government and international lecturers.

Example 10.2: One-Tailed Test

You want to know if male students spend more money on food in the cafe than female students do in a semester. The table below includes the mean, standard deviation, and sample size of each group. At a = 0.05 determine if the salaries are the same or not

Males	Females
$\bar{X}_1 = 3,200$	$\bar{X}_2 = 2,950$
$s_1 = 230$	$s_2 = 119$
$n_1 = 39$	$n_2 = 35$

Step 1: State hypotheses
H0 = There is no difference in money spent in the cafÃl' by gender
H1 = Men spend more than women in the cafe

Step 2: Set alpha and degrees of freedom
= 0.05, degrees of freedom = 39 + 35 - 2 = 72

Step 3: Find the t critical value
+1.67 one-tailed test.

Step 4: Compute value

$$t = \frac{(\bar{X}_1 - \bar{X}_2) - (\mu_1 - \mu_2)}{\sqrt{\frac{(n_1-1)s_1^2 + (n_2-1)s_2^2}{n_1 + n_2 - 2}} \sqrt{\frac{1}{n_1} + \frac{1}{n_2}}}$$

$$\frac{(3250 - 2950) - 0}{\sqrt{\frac{(39-1)230^2 + (35-1)119^2}{39 + 35 - 2}} \sqrt{\frac{1}{39} + \frac{1}{35}}} = 6.92$$

Step 5: Make Decision
Since the t-stat is greater than 1.67 we reject the null hypothesis

Step 6: Make Conclusion
There is evidence to consider that male students spend more than female students on food in the cafe in a semester.

Two Sample t-test Using R

Doing this in R is of course much easier. We are going to use the `ToothGrowth` data set and we want to see if there is a difference in tooth length by supplement type. Below is the code along with the results.

```
> t.test(formula=len~supp, data=ToothGrowth)

        Welch Two Sample t-test

data:  len by supp
t = 1.9153, df = 55.309, p-value = 0.06063
alternative hypothesis: true difference in means is not equal to 0
95 percent confidence interval:
 -0.1710156  7.5710156
sample estimates:
mean in group OJ mean in group VC
        20.66333         16.96333
```

To get these results you use the `t.test` function. Then you do the following...

1. You always place the continuous variable to the left (len) followed by the tilde

2. Type the categorical variable (supp).

3. Enter a comma

4. Put the name of the dataset.

The output indicated that there is no difference (t = 1.91 < 2.01) in the means of the tooth based on the supplement.

Paired T-Test

Paired sampling is used to compare two samples that are related to each other. Often the two samples come from the same source or person. Paired sample is commonly used in before and after experiments when you draw a sample from the same source before and after an intervention. For example, if you wanted to know if music affects reading comprehension you might measure a person's reading ability before listening to music and after to look for changes.

The primary number that needs to be calculated is the difference between the two scores. You need to know the following about the difference

CHAPTER 10. WHAT IS TWO SAMPLE HYPOTHESIS TESTING?

- The difference of the two samples

$$X_2 - X_1 = D$$

- The average amount of difference.

$$\bar{D} = \frac{\Sigma D}{n}$$

- The standard deviation of the difference

$$s_D = \sqrt{\frac{\Sigma D^2 - \frac{(\Sigma D)^2}{n}}{n-1}}$$

- Below is the equation for the t-stat

$$t = \frac{\bar{D} - \mu_D}{s_D/\sqrt{n}}$$

The steps for calculating a paired sample t-test is as follows

1. State hypotheses
2. Set level of significance
3. Determine degrees of freedom and critical value
4. Calculate t-test
5. Make decision
6. State the conclusion

Below is an example

Example 10.3

10 struggling students were put in study skills training. You want to see if study skills training had any effect on the students GPA. Below is a table that shows the students GPA before and after attending study skills training. Determine if the after GPA is greater than the before GPA at an alpha of 0.05.

Student	Before	After
1	2	3.5
2	3	3.8
3	2.5	3.7
4	3.1	4
5	1.9	1.8
6	1.7	1.6
7	2.3	3.4
8	3.2	2.3
9	2.3	2.3
10	1.9	3

Step 1: State hypotheses
H0: There is no difference in the before and after means
H1: There is a difference in the before and after means

Step 2: Set Significance
a = 0.05

Step 3: Find the t critical value
Df = n - 1 = 10 - 1 = 9
t critical = 1.83

Step 4: Compute value

CHAPTER 10. WHAT IS TWO SAMPLE HYPOTHESIS TESTING?

Student	Before (X_1)	After (X_2)	$D = X_2 - X_1$	$D^2 = (X_2 - X_1)^2$
1	2	3.5	1.5	2.25
2	3	3.8	0.8	0.64
3	2.5	3.7	1.2	1.44
4	3.1	4	0.9	0.81
5	1.9	1.8	-0.1	0.01
6	1.7	1.6	-0.1	0.01
7	2.3	3.4	1.1	1.21
8	3.2	2.3	-0.9	0.81
9	2.3	2.3	0	0
10	1.9	3	1.1	1.21
			$\sum D = 5.5$	$\sum D^2 = 8.39$

mean is

$$\bar{D} = \frac{\sum D}{n} = \frac{5.5}{10} = 0.55$$

standard deviation

$$s_D = \sqrt{\frac{\sum D^2 - \frac{(\sum D)^2}{n}}{n-1}} = \sqrt{\frac{8.39 - \frac{5.5^2}{10}}{10-1}} = 0.77$$

t stat is as follows

$$t = \frac{\bar{D} - \mu_D}{s_D/\sqrt{n}} = \frac{0.55 - 0}{0.77\sqrt{10}} = 2.29$$

Step 5: Make Decision
Since the t-stat is greater than the t critical value of 1.83, we reject the null hypothesis

Step 6: Make Conclusion
Since we reject the null hypothesis, we can state that the GPA after the study skills training is greater than before the training.

Paired t-test Using R

We will now do an example in R. We will use the `sleep` data and we want to determine if the use of a degree increase the amount of sleep people had. To do this we will use the extra variable in the `sleep` dataset.

If you look at the data set, you will see we need to make some adjustments in order to do the paired test. The first ten rows are the before data and the last ten rows are the after data. To use the function we need to use brackets next to the `sleep$extra` data to only use the rows we want. In addition, we want the last ten rows to be placed inside the function first because this is the same as the X_2 in our original equation. Lastly, we need to set the argument paired to TRUE. Below is the code with the output.

```
> t.test(sleep$extra[11:20], sleep$extra[1:10], paired=TRUE)

        Paired t-test

data:  sleep$extra[11:20] and sleep$extra[1:10]
t = -4.0621, df = 9, p-value = 0.002833
alternative hypothesis: true difference in means is not equal to 0
95 percent confidence interval:
 -2.4598858 -0.7001142
sample estimates:
mean of the differences
                   1.58
```

The test indicates that there is a difference and that people using the drug slept on average 1.6 hours longer.

Two-Sample Test of Proportions

Proportion tests are used when nominal data is involved. The goal is to determine if two parts of a whole are different from each other. The formula is below

$$\bar{p} = \frac{x_1+x_2}{n_1+n_2} \quad \bar{q} = 1-\bar{p} \quad \hat{p}_1 = \frac{x_1}{n_1} \quad \hat{p}_2 = \frac{x_2}{n_2}$$

$$t = \frac{(\hat{p}_1-\hat{p}_2)-(p_1-p_2)}{\sqrt{\bar{p}\bar{q}(\frac{1}{n_1}+\frac{1}{n_2})}}$$

Same steps as before for the other t-tests.

1. State hypotheses
2. Set level of significance
3. Determine degrees of freedom and critical value
4. Calculate t-test

5. Make decision

6. State the conclusion

Example 10.4

In a sample of 156 Freshmen 95 were living on-campus. In a sample of 123 sophomores 73 lived on campus. At an alpha of 0.05, determine if there is a difference in the proportion of people who lived on-campus.

Step 1: State hypotheses
H0 = The proportions are the same
H1 = The proportions are not the same

Step 2: Set alpha
a = 0.05

Step 3: Find the t critical value
±1.96 two-tailed test.

Step 4: Compute value

$$\bar{p} = \frac{95+73}{156+123} = 0.60 \quad \bar{q} = 1 - 0.60 = 0.40 \quad \hat{p}_1 = \frac{95}{156} = 0.61 \quad \hat{p}_2 = \frac{73}{123} = 0.59$$

$$t = \frac{(0.61-0.59)-0}{\sqrt{(0.60*0.40)(\frac{1}{156}+\frac{1}{123})}} = \frac{0.02}{0.05} = 0.4$$

Step 5: Make Decision
Since the t-stat value is 0.4 and this is less than z critical of 1.96 at a significance of 0.05 we do not reject the null hypothesis.

Step 6: Make Conclusion

Since we do reject the null we can conclude that there is no difference in the proportion of freshman and sophomore students who live on-campus.

Two-Sample Test of Proportion Using R

Below is the code to find the answer in R. We will use the `prop.test` function to calculate the values.

```
> prop.test(x = c(95, 73), n = c(156, 123))

    2-sample test for equality of proportions with
    continuity correction

data:  c(95, 73) out of c(156, 123)
X-squared = 0.019342, df = 1, p-value = 0.8894
alternative hypothesis: two.sided
95 percent confidence interval:
 -0.1075440  0.1385009
sample estimates:
   prop 1    prop 2
0.6089744 0.5934959
```

In order to do this, you have to have two values of x and this is done with the c function. This is also necessary for the sample size of n. Once the values are inputted, you get the results above.

Conclusion

In this chapter, we learned about comparing means from different populations. In most social science applications this is much more commonly used then the one sample test we learned in the previous chapter.

Points to Remember

- t-test can be used to compare means from different populations.

- A paired t-test is the standard for before and after experiments.

Exercises

1. The school is using a new reading software and wants to see if the software improves performance. Class 1, with 35 students uses the software while class 2 with 28 students does not. With an a = 0.05 determine if there is a difference in the mean of the reading grade of these two classes.

CHAPTER 10. WHAT IS TWO SAMPLE HYPOTHESIS TESTING?

Class 1	Class 2
$x_1 = 80$	$x_2 = 74$
$s_1 = 7.1$	$s_2 = 4.7$
$n_1 = 35$	$n_2 = 28$

2. The music department on-campus claims that people taking music courses get better grades than students who do not. A sample of 45 students taken music was taken as well as a sample of 33 students who are not in music classes. The GPA of the students are below. At an alpha of 0.05, determine if there is a different in the mean of the GPA of the two groups.

Music	Non-Music
$x_1 = 3.4$	$x_2 = 3.1$
$s_1 = 0.2$	$s_2 = 0.28$
$n_1 = 45$	$n_2 = 33$

3. Using R, determine if having an automatic transmission or not makes a difference in the miles per gallon in the mtcars dataset.

4. Using R, determine if having an automatic transmission or not makes a difference in the horsepower in the mtcars dataset.

5. The school decided to repeat the reading software experiment. This time the students reading scores were recorded before using the reading software and then after using the software. At a significance of 0.05, determine if there is a difference in the reading scores with the data below.

Student	Before	After
1	75	65
2	65	65
3	60	70
4	54	62
5	81	74
6	78	81
7	48	55
8	69	80
9	71	69
10	88	87

6. In a sample of 56 undergrads 23 were planning to work overseas. IN a sample of 70 sophomores, 40 were planning to work overseas. At an alpha of 0.05, determine if there is a difference in the proportion of undergrads that plan to work overseas.

7. Using R, in a sample of 62 women 19 were married. IN another sample of 34 women were 15 were married. At an alpha of 0.05, determine if there is a difference in the proportion of married women.

Chapter 11

What is Analysis of Variance (ANOVA)?

In this chapter, you will learn the following...

- How to calculate and interpret ANOVA results

Key Terms.
  ANOVA   Tukey Post-Hoc   F-distribution

Introduction

Analysis of variance (ANOVA) is used when you want to compared three or more sample means. There are several different versions of ANOVA. However, we will limit this chapter to the application of one-way ANOVA.

At this point, the statistical analysis is too complex to do by hand. Therefore, this chapter will only show you how to conduct the analysis in R. We will look at various terms and how to interpret them in the context of ANOVA but we will avoid equation and any hand calculation. In addition, instead of the standardized or t distribution, ANOVA uses the F distribution. Allows for degrees of freedom for both the numerator and the denominator. This is important because ANOVA requires this for obtaining the probabilities. You can see the appendix in order to see the f distribution.

ANOVA is often used for experiments to see if different treatments affect the means of subgroups on a given outcome variable. In addition, ANOVA is also used in situations in which an experiment may not be obvious. For example, if we want to compare the means of university students by class level we clearly have four distinct groups but we did not really do an experiment. Instead, what we did was allow the class level to be the treatment and see if this affects the mean of whatever we are measuring.

The Process

The steps for conducting ANOVA are mostly the same as for other hypothesis testing.

1. State the hypothesis

2. Set the significance

3. Calculate the means

4. Calculate degrees of freedom and the critical value

5. Compute

6. Make decision

7. If the null hypothesis is reject conduct a Tukey Post Hoc test

CHAPTER 11. WHAT IS ANALYSIS OF VARIANCE (ANOVA)?

8. Make conclusion

The actual steps are secondary to understanding the output in R. Below in Table 11.1 is an actual ANOVA table from R however, the format is not the same as found in R. Each box is labeled and defined below. The top row is information between groups while the bottom row is information within groups.

	Df	Sum Sq	Mean Sq	F value	Pr(>F)
a. variable	b. 3	c. 8.55	d. 2.85	e. 0.76	f. 0.533
g. Residuals	h. 16	i. 60.00	j. 3.75		

Table 11.1: ANOVA Results Example

a. This is the name of the grouping. It is a categorical independent variable.

b. Degrees of freedom for the categorical independent variable

c. Sum of squares between groups. The difference between the means of the different groups summed.

d. Mean square error between groups. Simply the average of the error between the groups remember error is any value different from the mean.

e. F value. This is the same as the t value in prior chapters. Depending on the significance level, you know if your results are in the critical rejection region in the distribution.

f. Pvalue. Here the results are not significant

g. Residuals is the within group error in the ANOVA table.

h. Degrees of freedom for the continuous dependent variable.

i. Sum of squares within the groups.

j. Mean of square within groups.

The most important value for someone new to statistics is the pvalue. This will tell you if at least one of the means is statistically significant from the others. Essentially what is happening with ANOVA is the same as with a t-test. You are comparing the mean of groups to the variability. The main difference is that the number of potential subgroups is 3 or more. We will now go through an example

Example 11.1

We want to determine if there is a difference in the Sepal.Length by Species in the iris dataset. We will use the aov function in R as well as the summary function to view the results. Below is the code.

```
> Example11.1<-aov(iris$Sepal.Length~iris$Species)
> summary(Example11.1)
              Df Sum Sq Mean Sq F value Pr(>F)
iris$Species   2  63.21  31.606   119.3 <2e-16 ***
Residuals    147  38.96   0.265
---
Signif. codes:  0 '***' 0.001 '**' 0.01 '*' 0.05 '.' 0.1 ' ' 1
```

We used the aov function and we inputted the formula. Everything was saved inside an object call Example 11.1. Then we used the summary function to get the results. You can see from the probability that the results are significant. Therefore, we know that the species have different Sepal lengths but we do not know exactly where the differences are. To determine this we have to do a Tukey Post Hoc test. This is done with the TukeyHSD function. Below is the code and results.

```
> TukeyHSD(Example11.1)
  Tukey multiple comparisons of means
    95% family-wise confidence level

Fit: aov(formula = iris$Sepal.Length ~ iris$Species, a = 0.1)

$`iris$Species`
                       diff       lwr       upr p adj
versicolor-setosa     0.930 0.6862273 1.1737727     0
virginica-setosa      1.582 1.3382273 1.8257727     0
virginica-versicolor  0.652 0.4082273 0.8957727     0
```

The focus for now is on the p adj column. If this is below 0.05 it means that there is a difference between the two Species that are in the same row. For example, if you subtract the mean of versicolor from setosa it comes to 0.93. Or in other words, the versicolar species has a sepal length that is 0.93 larger than setosa. The next two column are the confidence interval. We can confirm that veriscolor is larger then setosa by finding the mean of each species in the iris dataset. This requires some coding that is not a part of this course, however, it is listed below.

CHAPTER 11. WHAT IS ANALYSIS OF VARIANCE (ANOVA)?

```
> aggregate(iris$Sepal.Length,list(iris$Species),mean)
     Group.1     x
1     setosa  5.006
2 versicolor  5.936
3  virginica  6.588
```

If you subtract versicolor from setosa you will get the same value of 0.93. We could do this for the other two comparisons as well.

Conclusion

This chapter discussed the use of ANOVA. The important thing to remember is knowing to use ANOVA. ANOVA is used when you have three or more means that you want to compare.

Points to Remember

- ANOVA compares multiple means.
- The Tukey Post Hoc test is used if a difference in means is detected.

R Code Used

- aov(): ANOVA
- TukeyHSD(): Tukey post-hoc test

Exercises

1. Using R, use anova to determine if there is a difference in petal length by species in the iris dataset.

2. Using R, use anova to determine if there is a difference in sepal width by species in the iris dataset.

3. Using R, use anova to determine if there is a difference in petal width by species in the iris dataset.

Chapter 12

What is Correlation and Regression?

In this chapter, you will learn the following...

- The use of correlation.
- The purpose of regression.
- Interpreting regression results

Key Terms.
Correlation   Regression   Multiple Regression   R-Square

Introduction

In this chapter, we will look at correlation and regression. Correlation is a measure of the strength of the relationship between two or more continuous variables. An extension of correlation is regression. Regression is used to explain the variance of the dependent variable based on the influence or relationship of an independent variable. When one dependent and one independent variable is involved it is called simple regression when more than one independent variable is involved it is called multiple regression.

Multiple regression is a highly common statistical tool used in the social science and practically all researchers are exposed to it one way or another. Entire books are written on multiple regression. Therefore, we will only cover the most basic aspects of regression using R.

Scatter Plots and Correlation

One of the first things to do when looking at the relationship between two continuous variables is to examine a scatter plot as we learned in a prior chapter. The scatter plot allows you to see how the variables behave. Figure 12.1 shows different ways variables can interact in a scatter plot.

CHAPTER 12. WHAT IS CORRELATION AND REGRESSION?

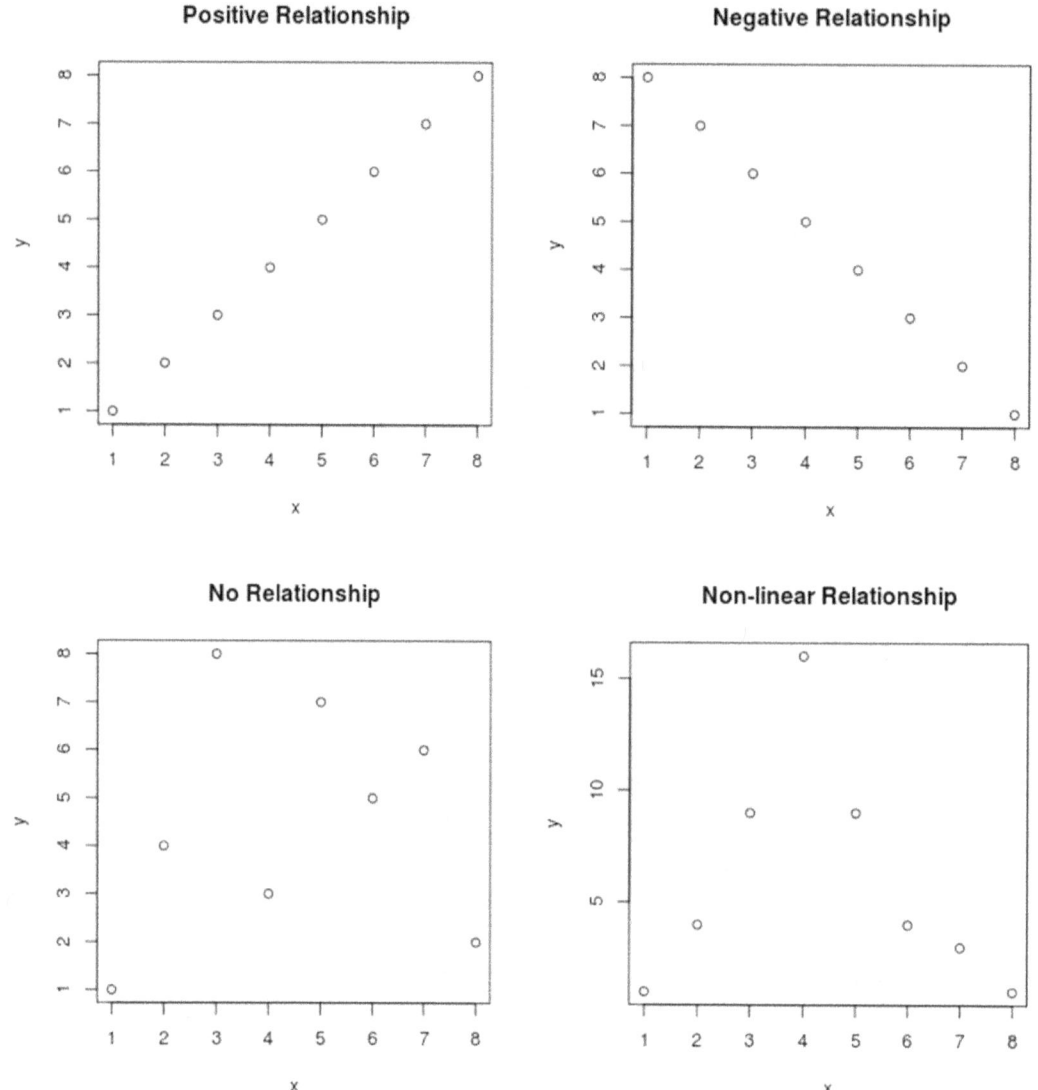

Figure 12.1: Relationship Types

As you look at each plot, you can see that the relationships in each are different. Sometimes the values increase together such as in the positive relationship chart. Other times one variable increase while the other decreases as in the negative relationship plot. Other times there is no visible relationship as in the no relationship plot. Lastly, sometimes the values increase together but then one variable starts to decrease while the other continues to increase as in the non-linear relationship plot.

However, it is not always wise to trust our eyes. There is a way to measure the strength of the linear relationship between two or more variables. The name of this statistical tool is called the Pearson's Product-moment correlation. The Pearson correlation uses a range of -1 to +1 to measure the strength of the relationship. A negative relationship means that the variables move in opposite directions. A positive relationship means that the variables move together in the same directions. Lastly, a low correlation means the variables do not move in any direction together.

There is an equation for calculating the correlation but we will use R to find the results. It is almost never practical to find the correlation by hand especially with real data because of the sample size. Figure 12.2 contains the same four plots we looked at previously but this time with the correlation included. The shorthand notation for correlation is r not to be confused with the software R.

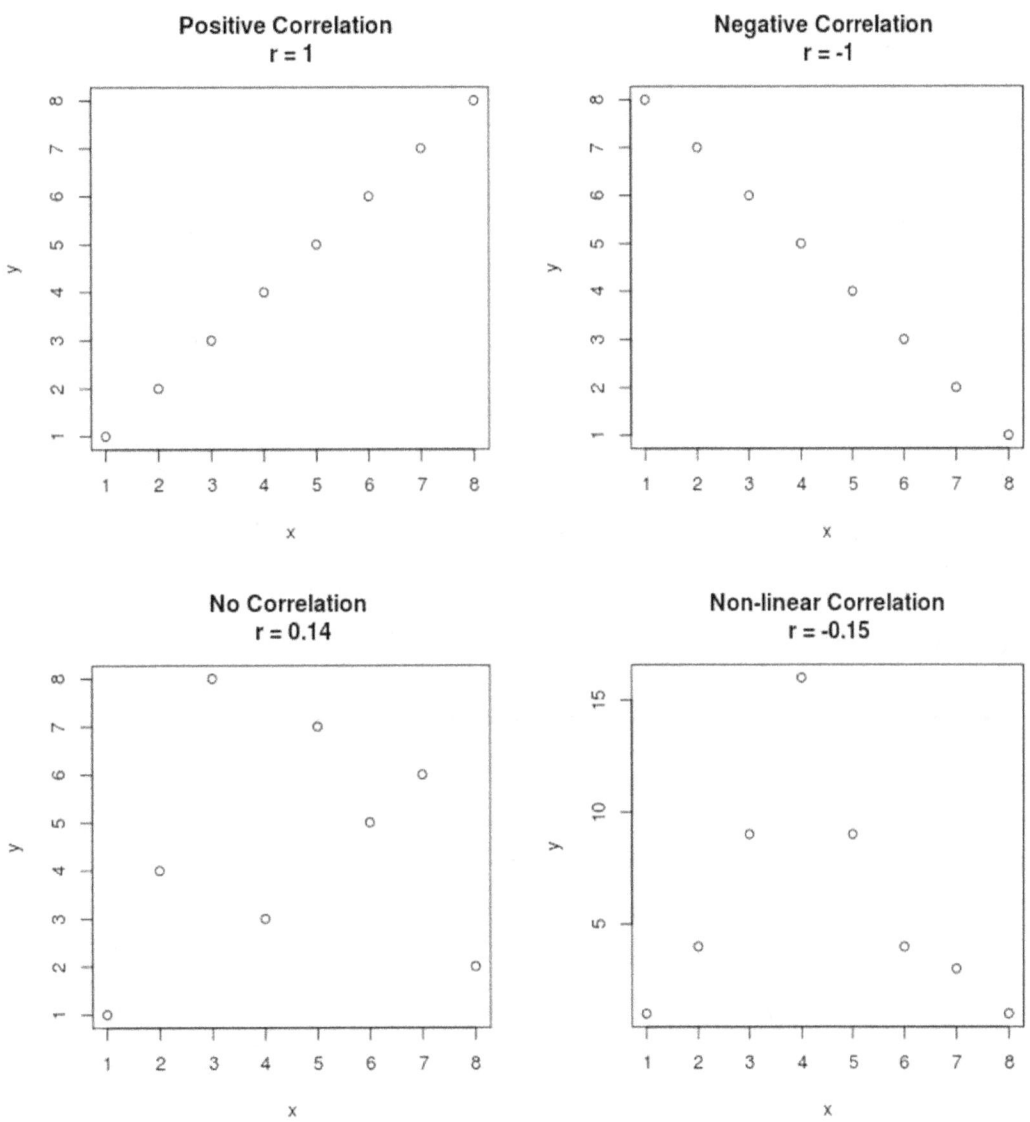

Figure 12.2: Relationship Types with Correlations

CHAPTER 12. WHAT IS CORRELATION AND REGRESSION?

A question to consider is when is a relationship strong when looking at the correlation. This is primarily a domain specific question. Different disciplines have different criteria for what a strong relationship is. The social sciences tend to be more tolerate and flexible than the hard sciences. Table 12.1 provides some generally guideline to use in social science disciplines.

Minimum		Maximum	Interpretation
-+ 0.01	to	-+ 0.20	Weak relationship
-+ 0.21	to	-+ 0.40	Slight relationship
-+ 0.41	to	-+ 0.70	Mild relationship
-+ 0.71	to	-+ 0.90	Strong relationship
-+ 0.91	to	-+ 1.00	Extremely Strong relationship

Table 12.1: Correlation Interpretation Guidelines

It is also possible to conduct a hypothesis test on a correlation. When you do this you are testing the null hypothesis that there is no relationship between the two variables. This will also be calculated using R. We will now turn our attention to an example in which we find the correlation between `Sepal.Length` and `Sepal.Width` in the iris dataset and conduct a hypothesis test.

Example 12.1

We want to find the correlation between Sepal.Length and Sepal.Width in the iris dataset below is the code.

```
> cor.test(iris$Sepal.Length,iris$Sepal.Width)

        Pearson's product-moment correlation

data:  iris$Sepal.Length and iris$Sepal.Width
t = -1.4403, df = 148, p-value = 0.1519
alternative hypothesis: true correlation is not equal to 0
95 percent confidence interval:
 -0.27269325  0.04351158
sample estimates:
       cor
-0.1175698
```

The p-value is above 0.05 which means there is no relationship between the two variables. In addition, the correlation was -0.11 which is a weak relationship. Lastly, all of this is confirm in Figure 12.3.

It is also possible to check all the correlations in your dataset at the same time. This can be done with the corr.test function from the psych package. You may need to install the psych package in order to use the corr.test function.

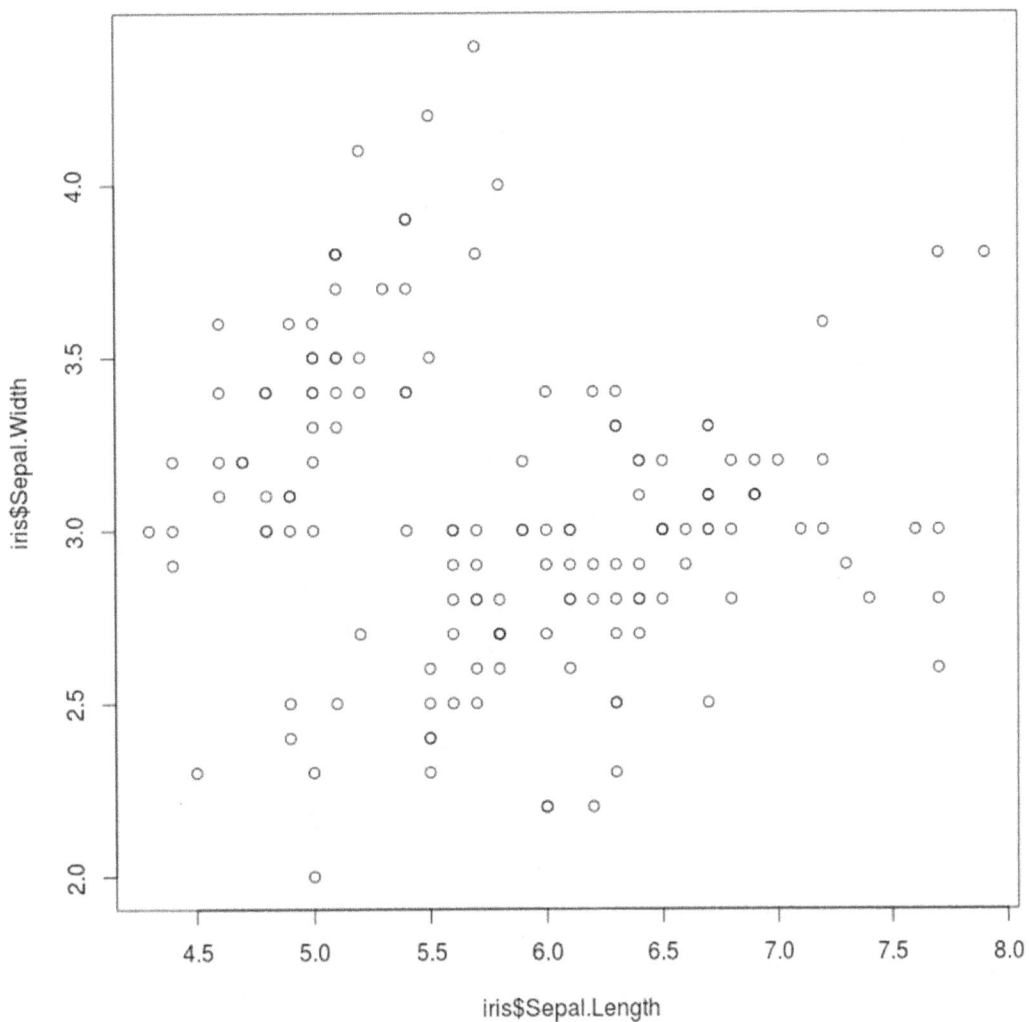

Figure 12.3: Scatter Plot of Example 12.1

Example 12.2

Find all the correlations and do the significance test for the continuous variables in the iris dataset. The code is below.

```
> corr.test(iris[,1:4])
Call:corr.test(x = iris[, 1:4])
Correlation matrix
             Sepal.Length Sepal.Width Petal.Length Petal.Width
Sepal.Length         1.00       -0.12         0.87        0.82
Sepal.Width         -0.12        1.00        -0.43       -0.37
Petal.Length         0.87       -0.43         1.00        0.96
Petal.Width          0.82       -0.37         0.96        1.00
Sample Size
[1] 150
Probability values (Entries above the diagonal are adjusted for
multiple tests.)
             Sepal.Length Sepal.Width Petal.Length Petal.Width
Sepal.Length         0.00        0.15            0           0
Sepal.Width          0.15        0.00            0           0
Petal.Length         0.00        0.00            0           0
Petal.Width          0.00        0.00            0           0
```

To see confidence intervals of the correlations, print with the short=FALSE option. In the code above, we used the brackets and put [,1:4] inside. This told R to use the first four columns or variables in the iris dataset. We only selected the first four columns because the fifth column is Species, which is a categorical variable. In the results, you can see all the correlations and the significance results. Notice we have the same non-significant relationship between Sepal.Length and Petal.Width

Simple Linear Regression

Linear regression is used to measure the dependence of a variable on one or more independent variables. The primary goal is to draw a line that reduces the amount of error and provides the best fit. We will look at same plots to make sense of this.

In Figure 12.4, we have X and Y. Our first line is a line that uses the mean of Y as the best line. You can see that this line has problems. First of all, none of the points are on the line this means that there is a lot of error or values different from the mean. The further a point is from the line the greater the error is. Secondly, the line ignores the upward trend of the data. This means the line is bias or not really listening to the data.

The problem with ignoring the trend is that it defeats the purpose of regression. There are two primary goals with regression.

1. To explain the relationship between variables

2. To predict future values

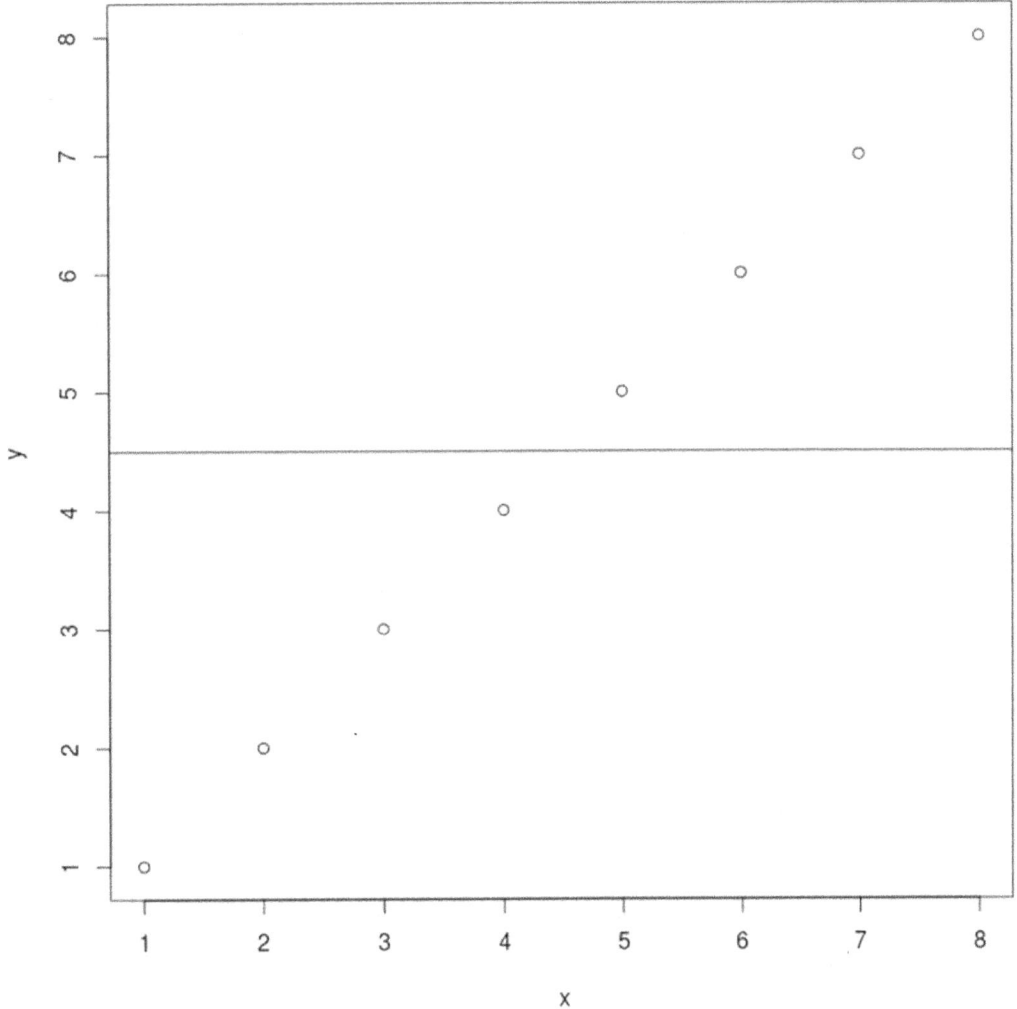

Figure 12.4: Scatter Plot with Mean Line

In Figure 12.4, you always have the same value of Y no matter what X is. Therefore, you are not explaining or predicting anything. Linear regression fixes these problems of using the mean by drawing a line that best accommodates both X and Y. In Figure 12.5, we now have two lines. The solid line is the mean and the dash line is the line using simple regression.

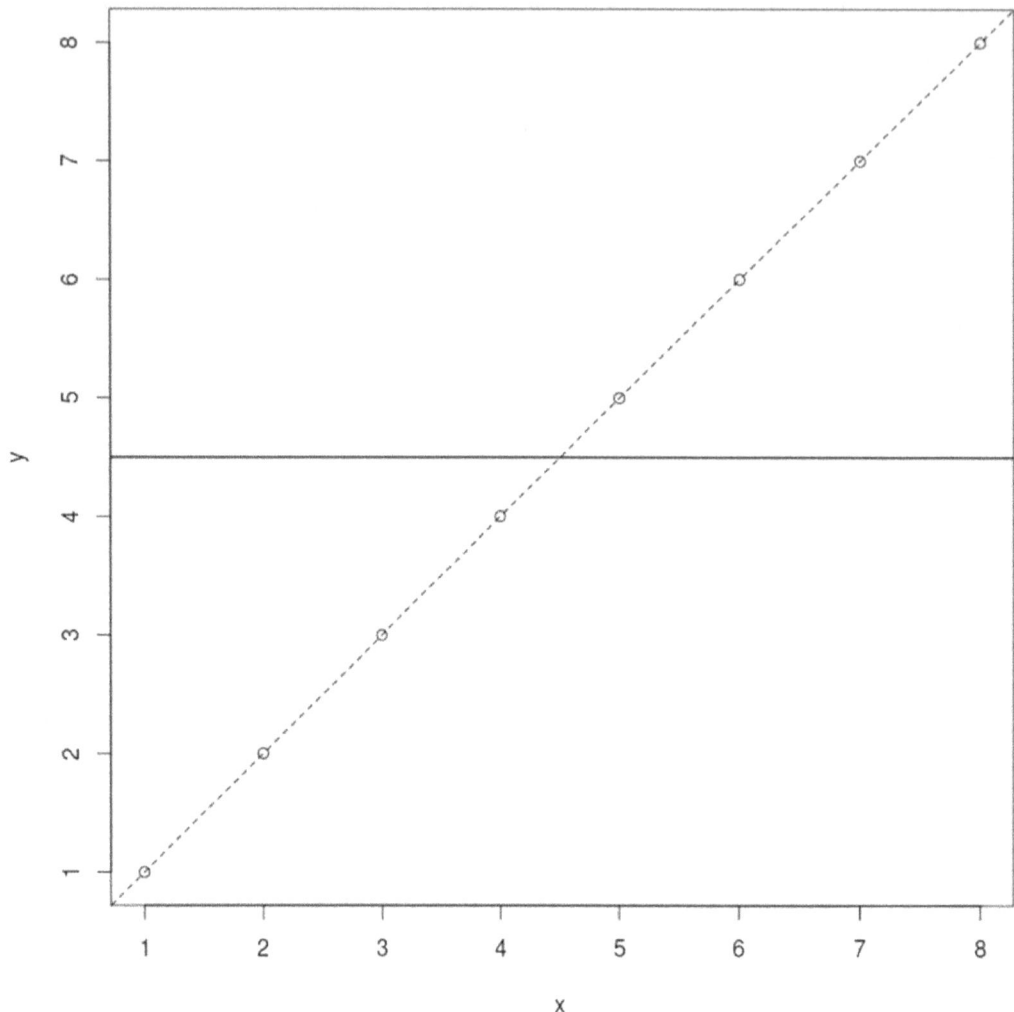

Figure 12.5: Scatter Plot with Mean Line & Regression Line

With the dashed lines, you can see clearly that all of the data points go through the line. This means that the regression line explains perfectly the values of Y when X is known. In fact, the measurement of the regression line is a value that tells you how much better the regression line explains the dependent variable when compared to the mean. Do not forget how the mean is used for many statistical calculations.

We will now go over how the regression line is created. Assuming you have both X and Y values, you need to find the following information.

- Slope of the line

- Y intercept

These concepts should have been covered in high school math but will be briefly defined. The slope is the steepness of the line of a linear equation. Calculating this by hand is not practical and we will allow R to do it, the y intercept is the place where the line crosses the y-axis and indicates when the x variable(s) is set to zero. The actually formula for a regression equation is below.

$$\hat{y} = b_1 X + b_0$$

$\hat{y}$ = predicted value of y
b_1 = slope of the regression line
b_0 = y intercept

Remember the primary goal of a regression line is to reduce the error between the X and Y values. This is done either to explain the Y variable or to predict future Y values. A measure of how good a regression line is the coefficient of determination also known as the r^2. Again, we will let R calculate this for us.

How good the r^2 needs to be depends on the discipline and context. A lower r^2 is more acceptable in social science then in hard science. In our previous example, the r^2 was 1, which is as perfect as the r^2 can be since its range is from 0-1.

It is important to understand that when you move towards using regression that you are beginning to make statistical models. A statistical model is trying to explain or predict values. This brings insights into decision-making and a is a core part of research.

We will now look at a real example from the iris data set.

Example 12.3

We are going to create a regression equation in which we use `Petal.Width` to explain `Sepal.Length` in the iris dataset.

In the code, we created an object called flowers1. In this object, we used the `lm` function to calculate our regression equation. We typed `Sepal.Length ~Petal.Width` and set the data argument to iris. When you press enter, nothing visible happens. We use the `summary` function to see the results.

```
flowers1<-lm(Sepal.Length~Petal.Width,data=iris)
summary(flowers1)
       ##
## Call:
## lm(formula = Sepal.Length ~ Petal.Width, data = iris)
##
## Residuals:
##     Min      1Q  Median      3Q     Max
## -1.38822 -0.29358 -0.04393  0.26429  1.34521
##
## Coefficients:
##              Estimate Std. Error t value Pr(>|t|)
## (Intercept)  4.77763    0.07293   65.51   <2e-16 ***
## Petal.Width  0.88858    0.05137   17.30   <2e-16 ***
## ---
## Signif. codes:  0 '***' 0.001 '**' 0.01 '*' 0.05 '.' 0.1 ' ' 1
##
## Residual standard error: 0.478 on 148 degrees of freedom
## Multiple R-squared:  0.669,  Adjusted R-squared:  0.6668
## F-statistic: 299.2 on 1 and 148 DF,  p-value: < 2.2e-16
```

There is a lot of information here. The residuals indicate the average distance from the actual data point to the regression line and indicate the minimum, quartiles, and maximum. The median value is -0.04, which is not bad in this context.

The intercept is 4.77, which means that when Petal.Width is zero Sepal.Length is 4.77. This does not make sense in this context because a flower always possesses both sepal length and petal width. However, there are times when the y-intercept is highly important in interpreting data. Underneath the intercept we have the independent variable and next is the coefficient.

The slope or the coefficient is 0.88 between Petal.Width and Sepal.Length. The Std.error is a measure of the difference in the error of the actual y values with the y values on the regression line. This is complex and beyond the scope of this book but is valuable when getting deeper into regression analysis.

After the standard errors we have the p-values, which are significant. The p-values become more important when we discussed multiple regression. Below this we have the r-square, which is 0.67. This means that Petal.Width explains 67% of the variance or deviation from the mean in Sepal.Length. Whether this is good or bad depends on context. Below this information is the F-stat, which should be familiar from our discussion on ANOVA. Regression using the same distribution to determine if the model is significant which in this case it is. If we wanted to write the regression equation it would look like the follow

$$Sepal.\hat{Length} = 0.88(Petal.width) + 4.77$$

What this means is that for every 1-unit increase in Petal.Width there is a .88 unit increase in Sepal.Length. Figure 12.6 is a visual of the regression model we have made.

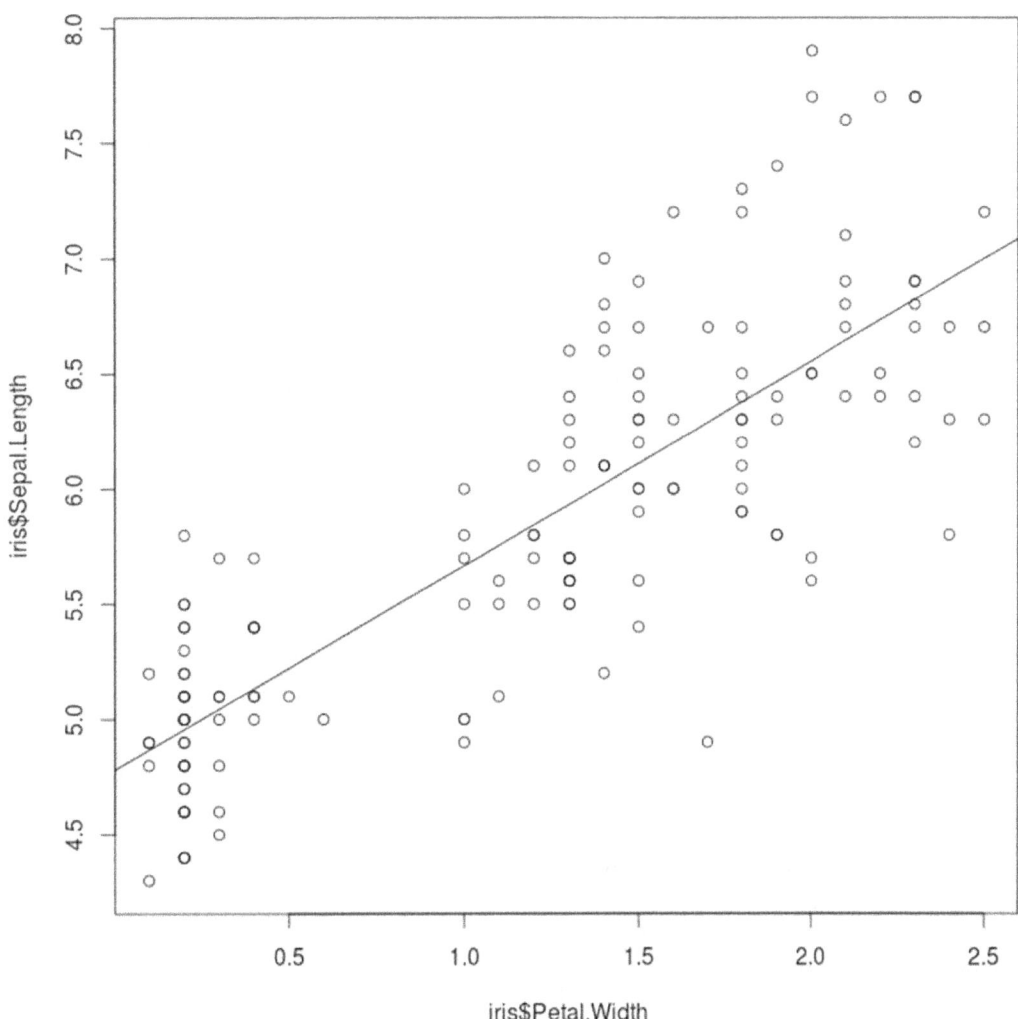

Figure 12.6: Simple Regression Plot

You can clearly see that many of the points are not on the line. This is why are r^2 is 0.67 and not 1. The line does a good job but is not perfect. This is how most regression models are.

Multiple Regression

Multiple regression involves more than one independent variable. This allows for a better understanding of the dependent variable through using more variables to explain it. In addition, using multiple variables helps to deal with correlations that look good but are not beneficial. A confounding variable is a variable that affects the relationship of other variables in the model.

For example, the relationship between shoe size and height. As people get taller their foot gets bigger. However, if you add age to the model the relationship between shoe size and height almost disappears because age is what is really affecting shoe size and height because as we get older (ie age) are height and shoe size change.

Example 12.4

We are going to explain Sepal.Length using the independent variables Petal.Width and Petal.Length. Below is the code and results.

```
flower2<-lm(Sepal.Length~Sepal.Width+Petal.Width,data=iris)
summary(flower2)
##
## Call:
## lm(formula = Sepal.Length ~ Sepal.Width + Petal.Width, data = iris)
##
## Residuals:
##     Min      1Q  Median      3Q     Max
## -1.2076 -0.2288 -0.0450  0.2266  1.1810
##
## Coefficients:
##             Estimate Std. Error t value Pr(>|t|)
## (Intercept)  3.45733    0.30919   11.18  < 2e-16 ***
## Sepal.Width  0.39907    0.09111    4.38 2.24e-05 ***
## Petal.Width  0.97213    0.05210   18.66  < 2e-16 ***
## ---
## Signif. codes:  0 '***' 0.001 '**' 0.01 '*' 0.05 '.' 0.1 ' ' 1
##
## Residual standard error: 0.4511 on 147 degrees of freedom
## Multiple R-squared:  0.7072, Adjusted R-squared:  0.7033
## F-statistic: 177.6 on 2 and 147 DF,  p-value: < 2.2e-16
```

The numbers are slightly different but the interpretation is the same. The only thing to mention is that Petal.Width has a slightly different value. This is because of the influence of Petal.Length, which also explains the dependent variable. The relationship is stronger because we are considering Sepal.Width and Petal.Length at the same time in the correlation. If we wanted to write the regression equation we would write the following...

$$Sepal.\hat{L}ength = 0.97(Petal.Width) + 0.40(Sepal.Width) + 4.77$$

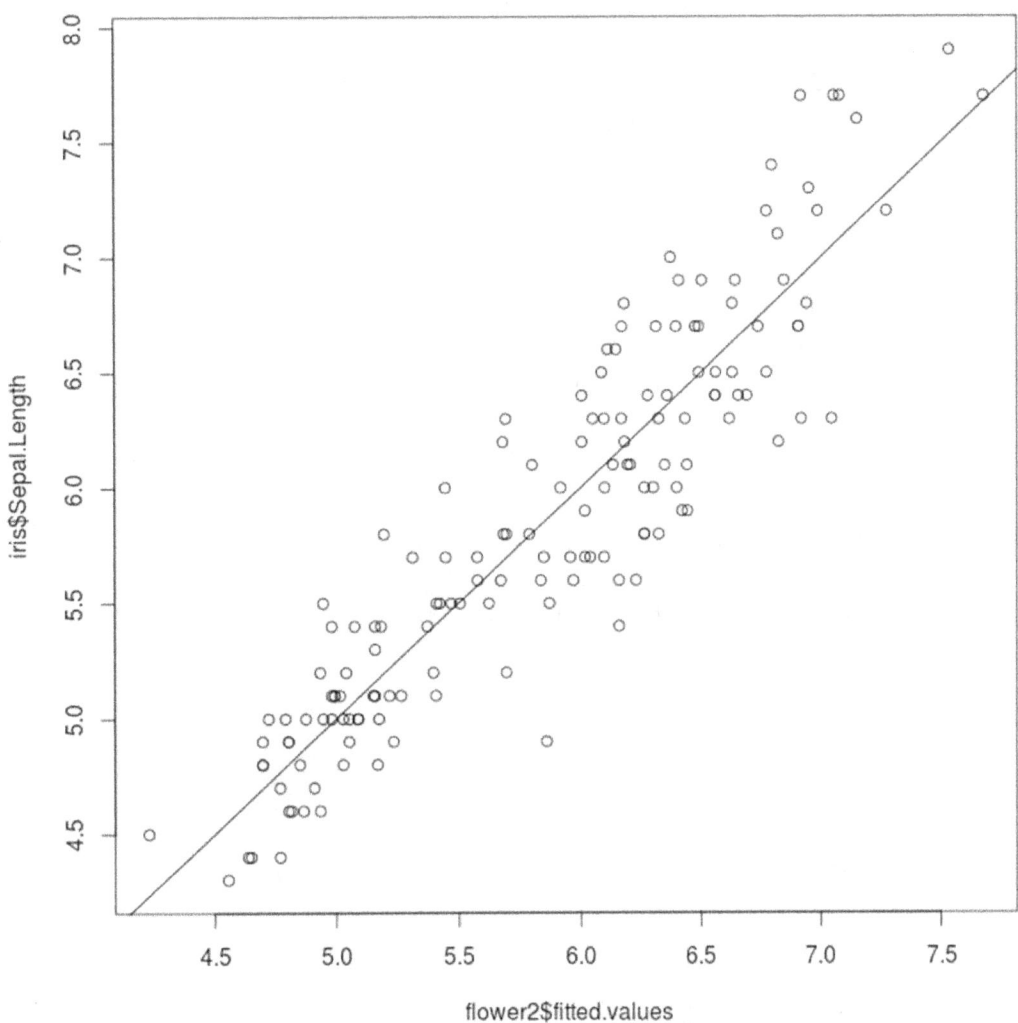

Figure 12.7: Multiple Regression Model

In English this means the following. When Petal.Width increases 1 unit Sepal.Length increase 0.97 when controlling for Sepal.Width. When Sepal.Width increases 1 unit Sepal.Length increases 0.4 units when controlling for Petal.Width. Figure 12.7 is a plot of the multiple regression model.

If you compare this plot with the plot using simple regression you can see that the points are closer to the line. This means that two variables are better than one in this case. Since the points are closer to the line there is less error in the results because the data points are closer to the predicted values on the line.

Figure 12.7 is not really a plot of the regression. Instead it is a comparison of the fitted and actual values. This plot was developed because multiple regression involves 3 or more variables. For each variable you must create an axis in the scatterplot. After 2 variables it becomes difficult to visualize therefore, it is common to plot the fitted values vs the actual values for visualization purposes.

What was explained here was the absolute minimum in understanding regression. There is much more to this than what we were able to cover.

Conclusion

Correlation and regression is the point at which statistics begins to truly get interesting. The principles laid down in this chapter can allow you to make powerful conclusion about data. Regression in particular is commonly used for making predictions about the world around us and serves as an entry point into machine learning and data science.

Points to Remember

- Correlation is a measure of the strength of the relationship between two or more variables

- Regression is used to predict or explain the influence of one or more variables on another.

R Code Used

- cor(): calculates correlation

- corr.test(): Calculates correlation and significance of many variables

- lm(): Creates regression model

Exercises

1. Find the correlation between ratings and complaints in the attitude dataset.

2. Find the correlation and significance test of all the variables in the mtcars dataset

3. Using simple regression to explain the dist variable with the speed variable in the cars dataset.

4. Using multiple regression to explain the mpg variable using all the other variables in the mtcars dataset.

Chapter 13

What is Chi-Square?

In this chapter, you will learn the following...

- Employ a chi-square test
- Interpret the results of a chi-square test

Key Terms.
Chi-Square   Goodness of Fit   Test of Independence

Introduction

Chi-square tests are another set of tools to assess data. The goodness of fit test tells us if our sampling distribution matches another theoretical distribution. The test of independence is useful for determining if there is a relationship between categorical variables. The chi-square test assumes a two-tail hypothesis test. In addition, the chi-square distribution is used and the shape of it changes depending on the degrees of freedom. See the appendix to understand more about the chi-square.

Goodness of Fit

The chi-square goodness of fit test is used to compare the sampling distribution to an expected probability distribution such as a normal distribution. The equation for calculating this is actually simple and we can calculate this for small sample sizes. Below is the equation.

$$x^2 = \sum \frac{(O-E)^2}{E}$$

O = Observed value
E = Expected value

The steps for completing this test are as follows...

- State the hypotheses
- Set significance level
- Determine the critical value
- Compute the chi0square
- Make decision
- Draw conclusion

Example 13.1

The cafe director wants to determine if the same number of students eat at the cafe every day. Data was collected for one week as shown below

Day	Monday	Tuesday	Wednesday	Thursday	Friday
Number of Students	68	80	64	56	52

Since there were five days. The cafe director expects the sample to be divided into five equal size groups each representing 20% of the data.

With an alpha of 0.05 determine if these values are the same.

Step1: Hypotheses
H0 = The proportions are all equal (0.2)
H1 = At least two of the proportions are not all equal (0.2)

Step 2: Set significance
Alpha = 0.05

Step 3: Find critical region
df = c - 1 = 5 - 1 = 4 chi-square = 9.49

Step 4: Compute the chi-square

O	P	E= np	O - E	$(O-E)^2$	$\frac{(O-E)^2}{E}$
68	0.2	320(0.20) = 64	4	16	0.25
80	0.2	320(0.20) = 64	16	256	4
64	0.2	320(0.20) = 64	0	0	0
56	0.2	320(0.20) = 64	-8	64	1
52	0.2	320(0.20) = 64	-12	144	2.25
Total = 320	1.00		0		$x^2 = \sum \frac{(O-E)^2}{E} = 7.75$

Step 5: Decisions
The computed chi-square is 7.75, which is less than the critical value of 9.49. Therefore, we do not reject the null hypothesis

Step 6: Conclusion
Since we do not reject the null we can see that the proportion of students who come to the cafe are equal.

Below is how to complete this analysis using R.

Find Goodness of Fit Using R

```
> chisq.test(x=c(68,80,64,56,52),p=c(.2,.2,.2,.2,.2))

    Chi-squared test for given probabilities

data:  c(68, 80, 64, 56, 52)
X-squared = 7.5, df = 4, p-value = 0.1117
```

In the R example, we have to set the value of x by including each value, we also must set the p for each value. This is important because the proportions do not have to be equal but can be set depending on the problem.

Test of Independence

The chi-square test of independence is used to determine if there is a relationship between nominal variables. The data is often in the form of a table. The actually calculation is a little more complex so we will use R for our example

Example 13.2

We will use the Computers dataset from the Ecdat package and see if there is a difference between the categorical variables multi and cd.

First, we will make a contingency table of the two variables. A contingency table tells you the values of the two variables when they are both present in the sample

```
> table(Computers$multi,Computers$cd)

      no   yes
  no  3351 2035
  yes    0  873
```

We now have an idea of the relationship between the two variables. Now we while calculate the chi-square with the chisq.test function.

```
> chisq.test(Computers$multi,Computers$cd)

    Pearson's Chi-squared test with Yates' continuity correction

data:  Computers$multi and Computers$cd
X-squared = 1166.5, df = 1, p-value < 2.2e-16
```

The pvalue is less than 0.05. This means that results indicate there is relationship between the two variables.

Conclusion

The chi-square test will become much more important as you journey into more advance statistical analysis. This chapter provided an exposure to the ideas related to this statistical test.

Points to Remember

- Chi-square determines relationships between categorical variables.

- Chi-square is also used fo comparing distributions.

R Code Used

- chisq.test(): Calculates chi-square results

Exercises

1. The administration wants to determine if the same assembly attendance is consistent each week. Data was collected for eight weeks as shown below

1	2	3	4	5	6	7	8
680	680	800	740	640	560	520	670

   With an alpha of 0.05 determine if these values are the same

2. Using R, Use the occupationalStatus dataset in R and determine if there is a relationship between origin and destination

APA for Statistical Results

Mean and Standard Deviation

Mean and Standard Deviation are often presented in parentheses. M stands for mean and SD stands for standard deviation. Both M and SD are in italics. Below is an example

- The sample as a whole disagree with the statements about cafÃl' food ($M = 1.22$, $SD = 0.15$)

Percentages

Percentages are displayed in parentheses with no decimal places:

- Nearly half (49%) of the sample were women.

T-Test

T Tests report the degrees of freedom in parentheses. Following that, report the t statistic (rounded to two decimal places), and the significance level. The t and the p are in italics.

- There was a significant effect for gender, $t(54) = 2.43$, $p < .05$, with men receiving higher scores than women.

ANOVA

ANOVAs (both one-way and two-way) are reported like the t test, but there are two degrees-of-freedom numbers to report. First you share the between-groups degrees of freedom, then report the within-groups degrees of freedom (separated by a comma). Lastly, report the F statistic and the significance level.

- There was a significant main effect for treatment, $F(1, 76) = 2.37$, $p < .05$.

Correlation

Correlations are reported with the degrees of freedom (which is n-2) in parentheses and the significance level:

- The two variables were moderately correlated, $r(55) = .49$, $p < .05$.

Regression

Regression results are often best presented in a table. In text, present the slope (beta), along with the t-test and the significance level. Lastly, share the r^2 results

- Sleep quality significantly predicted exam scores, $b = -.34$, $t(225) = 4.53$, $p < .05$. Sleep quality also explained a significant proportion of variance in depression scores, $r^2 = .12$, $F(1, 205) = 12.34$, $p < .05$.

Chi-Square

Chi-Square statistics are reported with degrees of freedom and sample size in the parentheses, the chi-square value (rounded to two decimal places), and the significance level:

- The percentage of participants that were students did not differ by gender, $x^2(1, n = 40) = 0.79$, $p = .35$.

When in doubt all statistical short hand terms f, m, n, p, r, sd, t, x^2 should be in italics.

Appendix A

Appendix

STUDENT'S t PERCENTAGE POINTS

v	60.0%	66.7%	75.0%	80.0%	87.5%	90.0%	95.0%	97.5%	99.0%	99.5%	99.9%
1	0.325	0.577	1.000	1.376	2.414	3.078	6.314	12.706	31.821	63.657	318.31
2	0.289	0.500	0.816	1.061	1.604	1.886	2.920	4.303	6.965	9.925	22.327
3	0.277	0.476	0.765	0.978	1.423	1.638	2.353	3.182	4.541	5.841	10.215
4	0.271	0.464	0.741	0.941	1.344	1.533	2.132	2.776	3.747	4.604	7.173
5	0.267	0.457	0.727	0.920	1.301	1.476	2.015	2.571	3.365	4.032	5.893
6	0.265	0.453	0.718	0.906	1.273	1.440	1.943	2.447	3.143	3.707	5.208
7	0.263	0.449	0.711	0.896	1.254	1.415	1.895	2.365	2.998	3.499	4.785
8	0.262	0.447	0.706	0.889	1.240	1.397	1.860	2.306	2.896	3.355	4.501
9	0.261	0.445	0.703	0.883	1.230	1.383	1.833	2.262	2.821	3.250	4.297
10	0.260	0.444	0.700	0.879	1.221	1.372	1.812	2.228	2.764	3.169	4.144
11	0.260	0.443	0.697	0.876	1.214	1.363	1.796	2.201	2.718	3.106	4.025
12	0.259	0.442	0.695	0.873	1.209	1.356	1.782	2.179	2.681	3.055	3.930
13	0.259	0.441	0.694	0.870	1.204	1.350	1.771	2.160	2.650	3.012	3.852
14	0.258	0.440	0.692	0.868	1.200	1.345	1.761	2.145	2.624	2.977	3.787
15	0.258	0.439	0.691	0.866	1.197	1.341	1.753	2.131	2.602	2.947	3.733
16	0.258	0.439	0.690	0.865	1.194	1.337	1.746	2.120	2.583	2.921	3.686
17	0.257	0.438	0.689	0.863	1.191	1.333	1.740	2.110	2.567	2.898	3.646
18	0.257	0.438	0.688	0.862	1.189	1.330	1.734	2.101	2.552	2.878	3.610
19	0.257	0.438	0.688	0.861	1.187	1.328	1.729	2.093	2.539	2.861	3.579
20	0.257	0.437	0.687	0.860	1.185	1.325	1.725	2.086	2.528	2.845	3.552
21	0.257	0.437	0.686	0.859	1.183	1.323	1.721	2.080	2.518	2.831	3.527
22	0.256	0.437	0.686	0.858	1.182	1.321	1.717	2.074	2.508	2.819	3.505
23	0.256	0.436	0.685	0.858	1.180	1.319	1.714	2.069	2.500	2.807	3.485
24	0.256	0.436	0.685	0.857	1.179	1.318	1.711	2.064	2.492	2.797	3.467
25	0.256	0.436	0.684	0.856	1.178	1.316	1.708	2.060	2.485	2.787	3.450
26	0.256	0.436	0.684	0.856	1.177	1.315	1.706	2.056	2.479	2.779	3.435
27	0.256	0.435	0.684	0.855	1.176	1.314	1.703	2.052	2.473	2.771	3.421
28	0.256	0.435	0.683	0.855	1.175	1.313	1.701	2.048	2.467	2.763	3.408
29	0.256	0.435	0.683	0.854	1.174	1.311	1.699	2.045	2.462	2.756	3.396
30	0.256	0.435	0.683	0.854	1.173	1.310	1.697	2.042	2.457	2.750	3.385
35	0.255	0.434	0.682	0.852	1.170	1.306	1.690	2.030	2.438	2.724	3.340
40	0.255	0.434	0.681	0.851	1.167	1.303	1.684	2.021	2.423	2.704	3.307
45	0.255	0.434	0.680	0.850	1.165	1.301	1.679	2.014	2.412	2.690	3.281
50	0.255	0.433	0.679	0.849	1.164	1.299	1.676	2.009	2.403	2.678	3.261
55	0.255	0.433	0.679	0.848	1.163	1.297	1.673	2.004	2.396	2.668	3.245
60	0.254	0.433	0.679	0.848	1.162	1.296	1.671	2.000	2.390	2.660	3.232
∞	0.253	0.431	0.674	0.842	1.150	1.282	1.645	1.960	2.326	2.576	3.090

	0	0.01	0.02	0.03	0.04	0.05	0.06	0.07	0.08	0.09
0	0.5000	0.5040	0.5080	0.5120	0.5160	0.5199	0.5239	0.5279	0.5319	0.5359
0.1	0.5398	0.5438	0.5478	0.5517	0.5557	0.5596	0.5636	0.5675	0.5714	0.5753
0.2	0.5793	0.5832	0.5871	0.5910	0.5948	0.5987	0.6026	0.6064	0.6103	0.6141
0.3	0.6179	0.6217	0.6255	0.6293	0.6331	0.6368	0.6406	0.6443	0.6480	0.6517
0.4	0.6554	0.6591	0.6628	0.6664	0.6700	0.6736	0.6772	0.6808	0.6844	0.6879
0.5	0.6915	0.6950	0.6985	0.7019	0.7054	0.7088	0.7123	0.7157	0.7190	0.7224
0.6	0.7257	0.7291	0.7324	0.7357	0.7389	0.7422	0.7454	0.7486	0.7517	0.7549
0.7	0.7580	0.7611	0.7642	0.7673	0.7704	0.7734	0.7764	0.7794	0.7823	0.7852
0.8	0.7881	0.7910	0.7939	0.7967	0.7995	0.8023	0.8051	0.8078	0.8106	0.8133
0.9	0.8159	0.8186	0.8212	0.8238	0.8264	0.8289	0.8315	0.8340	0.8365	0.8389
1	0.8413	0.8438	0.8461	0.8485	0.8508	0.8531	0.8554	0.8577	0.8599	0.8621
1.1	0.8643	0.8665	0.8686	0.8708	0.8729	0.8749	0.8770	0.8790	0.8810	0.8830
1.2	0.8849	0.8869	0.8888	0.8907	0.8925	0.8944	0.8962	0.8980	0.8997	0.9015
1.3	0.9032	0.9049	0.9066	0.9082	0.9099	0.9115	0.9131	0.9147	0.9162	0.9177
1.4	0.9192	0.9207	0.9222	0.9236	0.9251	0.9265	0.9279	0.9292	0.9306	0.9319
1.5	0.9332	0.9345	0.9357	0.9370	0.9382	0.9394	0.9406	0.9418	0.9429	0.9441
1.6	0.9452	0.9463	0.9474	0.9484	0.9495	0.9505	0.9515	0.9525	0.9535	0.9545
1.7	0.9554	0.9564	0.9573	0.9582	0.9591	0.9599	0.9608	0.9616	0.9625	0.9633
1.8	0.9641	0.9649	0.9656	0.9664	0.9671	0.9678	0.9686	0.9693	0.9699	0.9706
1.9	0.9713	0.9719	0.9726	0.9732	0.9738	0.9744	0.9750	0.9756	0.9761	0.9767
2	0.9772	0.9778	0.9783	0.9788	0.9793	0.9798	0.9803	0.9808	0.9812	0.9817
2.1	0.9821	0.9826	0.9830	0.9834	0.9838	0.9842	0.9846	0.9850	0.9854	0.9857
2.2	0.9861	0.9864	0.9868	0.9871	0.9875	0.9878	0.9881	0.9884	0.9887	0.9890
2.3	0.9893	0.9896	0.9898	0.9901	0.9904	0.9906	0.9909	0.9911	0.9913	0.9916
2.4	0.9918	0.9920	0.9922	0.9925	0.9927	0.9929	0.9931	0.9932	0.9934	0.9936
2.5	0.9938	0.9940	0.9941	0.9943	0.9945	0.9946	0.9948	0.9949	0.9951	0.9952
2.6	0.9953	0.9955	0.9956	0.9957	0.9959	0.9960	0.9961	0.9962	0.9963	0.9964
2.7	0.9965	0.9966	0.9967	0.9968	0.9969	0.9970	0.9971	0.9972	0.9973	0.9974
2.8	0.9974	0.9975	0.9976	0.9977	0.9977	0.9978	0.9979	0.9979	0.9980	0.9981
2.9	0.9981	0.9982	0.9982	0.9983	0.9984	0.9984	0.9985	0.9985	0.9986	0.9986
3	0.9987	0.9987	0.9987	0.9988	0.9988	0.9989	0.9989	0.9989	0.9990	0.9990

PERCENTAGE POINTS OF THE F DISTRIBUTION

$v_2 \backslash v_1$	q	2	3	4	5	6	7	8	10	12	15	20	30	50	∞
1	0.500	1.50	1.71	1.82	1.89	1.94	1.98	2.00	2.04	2.07	2.09	2.12	2.15	2.17	2.20
	0.600	2.63	2.93	3.09	3.20	3.27	3.32	3.36	3.41	3.45	3.48	3.52	3.56	3.59	3.64
	0.667	4.00	4.42	4.64	4.78	4.88	4.95	5.00	5.08	5.13	5.18	5.24	5.29	5.33	5.39
	0.750	7.50	8.20	8.58	8.82	8.98	9.10	9.19	9.32	9.41	9.50	9.58	9.67	9.74	9.85
	0.800	12.0	13.1	13.6	14.0	14.3	14.4	14.6	14.8	14.9	15.0	15.2	15.3	15.4	15.6
2	0.500	1.00	1.13	1.21	1.25	1.28	1.30	1.32	1.35	1.36	1.38	1.39	1.41	1.42	1.44
	0.600	1.50	1.64	1.72	1.76	1.80	1.82	1.84	1.86	1.88	1.89	1.91	1.92	1.94	1.96
	0.667	2.00	2.15	2.22	2.27	2.30	2.33	2.34	2.37	2.38	2.40	2.42	2.43	2.45	2.47
	0.750	3.00	3.15	3.23	3.28	3.31	3.34	3.35	3.38	3.39	3.41	3.43	3.44	3.46	3.48
	0.800	4.00	4.16	4.24	4.28	4.32	4.34	4.36	4.38	4.40	4.42	4.43	4.45	4.47	4.48
3	0.500	0.88	1.00	1.06	1.10	1.13	1.15	1.16	1.18	1.20	1.21	1.23	1.24	1.25	1.27
	0.600	1.26	1.37	1.43	1.47	1.49	1.51	1.52	1.54	1.55	1.56	1.57	1.58	1.59	1.60
	0.667	1.62	1.72	1.77	1.80	1.82	1.83	1.84	1.86	1.87	1.88	1.89	1.90	1.90	1.91
	0.750	2.28	2.36	2.39	2.41	2.42	2.43	2.44	2.44	2.45	2.46	2.46	2.47	2.47	2.47
	0.800	2.89	2.94	2.96	2.97	2.97	2.97	2.98	2.98	2.98	2.98	2.98	2.98	2.98	2.98
4	0.500	0.83	0.94	1.00	1.04	1.06	1.08	1.09	1.11	1.13	1.14	1.15	1.16	1.18	1.19
	0.600	1.16	1.26	1.31	1.34	1.36	1.37	1.38	1.40	1.41	1.42	1.43	1.43	1.44	1.45
	0.667	1.46	1.55	1.58	1.61	1.62	1.63	1.64	1.65	1.65	1.66	1.67	1.67	1.68	1.68
	0.750	2.00	2.05	2.06	2.07	2.08	2.08	2.08	2.08	2.08	2.08	2.08	2.08	2.08	2.08
	0.800	2.47	2.48	2.48	2.48	2.47	2.47	2.47	2.46	2.46	2.45	2.44	2.44	2.43	2.43
5	0.500	0.80	0.91	0.96	1.00	1.02	1.04	1.05	1.07	1.09	1.10	1.11	1.12	1.13	1.15
	0.600	1.11	1.20	1.24	1.27	1.29	1.30	1.31	1.32	1.33	1.34	1.34	1.35	1.36	1.37
	0.667	1.38	1.45	1.48	1.50	1.51	1.52	1.53	1.53	1.54	1.54	1.54	1.55	1.55	1.55
	0.750	1.85	1.88	1.89	1.89	1.89	1.89	1.89	1.89	1.89	1.89	1.88	1.88	1.88	1.87
	0.800	2.26	2.25	2.24	2.23	2.22	2.21	2.20	2.19	2.18	2.18	2.17	2.16	2.15	2.13
6	0.500	0.78	0.89	0.94	0.98	1.00	1.02	1.03	1.05	1.06	1.07	1.08	1.10	1.11	1.12
	0.600	1.07	1.16	1.20	1.22	1.24	1.25	1.26	1.27	1.28	1.29	1.29	1.30	1.31	1.31
	0.667	1.33	1.39	1.42	1.44	1.44	1.45	1.45	1.46	1.46	1.47	1.47	1.47	1.47	1.47
	0.750	1.76	1.78	1.79	1.79	1.78	1.78	1.78	1.77	1.77	1.76	1.76	1.75	1.75	1.74
	0.800	2.13	2.11	2.09	2.08	2.06	2.05	2.04	2.03	2.02	2.01	2.00	1.98	1.97	1.95
7	0.500	0.77	0.87	0.93	0.96	0.98	1.00	1.01	1.03	1.04	1.05	1.07	1.08	1.09	1.10
	0.600	1.05	1.13	1.17	1.19	1.21	1.22	1.23	1.24	1.24	1.25	1.26	1.26	1.27	1.27
	0.667	1.29	1.35	1.38	1.39	1.40	1.40	1.41	1.41	1.41	1.41	1.41	1.42	1.42	1.42
	0.750	1.70	1.72	1.72	1.71	1.71	1.70	1.70	1.69	1.68	1.68	1.67	1.66	1.66	1.65
	0.800	2.04	2.02	1.99	1.97	1.96	1.94	1.93	1.92	1.91	1.89	1.88	1.86	1.85	1.83
8	0.500	0.76	0.86	0.91	0.95	0.97	0.99	1.00	1.02	1.03	1.04	1.05	1.07	1.07	1.09
	0.600	1.03	1.11	1.15	1.17	1.19	1.20	1.20	1.21	1.22	1.22	1.23	1.24	1.24	1.25
	0.667	1.26	1.32	1.35	1.36	1.36	1.37	1.37	1.37	1.37	1.38	1.38	1.38	1.37	1.37
	0.750	1.66	1.67	1.66	1.66	1.65	1.64	1.64	1.63	1.62	1.62	1.61	1.60	1.59	1.58
	0.800	1.98	1.95	1.92	1.90	1.88	1.87	1.86	1.84	1.83	1.81	1.80	1.78	1.76	1.74

APPENDIX A. APPENDIX

PERCENTAGE POINTS OF THE *F* DISTRIBUTION

$v_2 \backslash v_1$	q	2	3	4	5	6	7	8	10	12	15	20	30	50	∞
9	0.500	0.75	0.85	0.91	0.94	0.96	0.98	0.99	1.01	1.02	1.03	1.04	1.05	1.06	1.08
	0.600	1.02	1.10	1.13	1.15	1.17	1.18	1.18	1.19	1.20	1.21	1.21	1.22	1.22	1.22
	0.667	1.24	1.30	1.32	1.33	1.34	1.34	1.34	1.34	1.35	1.35	1.35	1.34	1.34	1.34
	0.750	1.62	1.63	1.63	1.62	1.61	1.60	1.60	1.59	1.58	1.57	1.56	1.55	1.54	1.53
	0.800	1.93	1.90	1.87	1.85	1.83	1.81	1.80	1.78	1.76	1.75	1.73	1.71	1.70	1.67
10	0.500	0.74	0.85	0.90	0.93	0.95	0.97	0.98	1.00	1.01	1.02	1.03	1.05	1.06	1.07
	0.600	1.01	1.08	1.12	1.14	1.15	1.16	1.17	1.18	1.18	1.19	1.19	1.20	1.20	1.21
	0.667	1.23	1.28	1.30	1.31	1.32	1.32	1.32	1.32	1.32	1.32	1.32	1.32	1.32	1.31
	0.750	1.60	1.60	1.59	1.59	1.58	1.57	1.56	1.55	1.54	1.53	1.52	1.51	1.50	1.48
	0.800	1.90	1.86	1.83	1.80	1.78	1.77	1.75	1.73	1.72	1.70	1.68	1.66	1.65	1.62
11	0.500	0.74	0.84	0.89	0.93	0.95	0.96	0.98	0.99	1.01	1.02	1.03	1.04	1.05	1.06
	0.600	1.00	1.07	1.11	1.13	1.14	1.15	1.16	1.17	1.17	1.18	1.18	1.18	1.19	1.19
	0.667	1.22	1.27	1.29	1.30	1.30	1.30	1.30	1.30	1.30	1.30	1.30	1.30	1.30	1.29
	0.750	1.58	1.58	1.57	1.56	1.55	1.54	1.53	1.52	1.51	1.50	1.49	1.48	1.47	1.45
	0.800	1.87	1.83	1.80	1.77	1.75	1.73	1.72	1.69	1.68	1.66	1.64	1.62	1.60	1.57
12	0.500	0.73	0.84	0.89	0.92	0.94	0.96	0.97	0.99	1.00	1.01	1.02	1.03	1.04	1.06
	0.600	0.99	1.07	1.10	1.12	1.13	1.14	1.15	1.16	1.16	1.17	1.17	1.17	1.18	1.18
	0.667	1.21	1.26	1.27	1.28	1.29	1.29	1.29	1.29	1.29	1.29	1.29	1.28	1.28	1.27
	0.750	1.56	1.56	1.55	1.54	1.53	1.52	1.51	1.50	1.49	1.48	1.47	1.45	1.44	1.42
	0.800	1.85	1.80	1.77	1.74	1.72	1.70	1.69	1.66	1.65	1.63	1.61	1.59	1.57	1.54
13	0.500	0.73	0.83	0.88	0.92	0.94	0.96	0.97	0.98	1.00	1.01	1.02	1.03	1.04	1.05
	0.600	0.98	1.06	1.09	1.11	1.13	1.13	1.14	1.15	1.15	1.16	1.16	1.16	1.17	1.17
	0.667	1.20	1.25	1.26	1.27	1.28	1.28	1.28	1.28	1.28	1.28	1.27	1.27	1.27	1.26
	0.750	1.55	1.55	1.53	1.52	1.51	1.50	1.49	1.48	1.47	1.46	1.45	1.43	1.42	1.40
	0.800	1.83	1.78	1.75	1.72	1.69	1.68	1.66	1.64	1.62	1.60	1.58	1.56	1.54	1.51
14	0.500	0.73	0.83	0.88	0.91	0.94	0.95	0.96	0.98	0.99	1.00	1.01	1.03	1.04	1.05
	0.600	0.98	1.05	1.09	1.11	1.12	1.13	1.13	1.14	1.14	1.15	1.15	1.16	1.16	1.16
	0.667	1.19	1.24	1.26	1.26	1.27	1.27	1.27	1.27	1.27	1.26	1.26	1.26	1.25	1.24
	0.750	1.53	1.53	1.52	1.51	1.50	1.49	1.48	1.46	1.45	1.44	1.43	1.41	1.40	1.38
	0.800	1.81	1.76	1.73	1.70	1.67	1.65	1.64	1.62	1.60	1.58	1.56	1.53	1.51	1.48
15	0.500	0.73	0.83	0.88	0.91	0.93	0.95	0.96	0.98	0.99	1.00	1.01	1.02	1.03	1.05
	0.600	0.97	1.05	1.08	1.10	1.11	1.12	1.13	1.13	1.14	1.14	1.15	1.15	1.15	1.15
	0.667	1.18	1.23	1.25	1.25	1.26	1.26	1.26	1.26	1.26	1.25	1.25	1.25	1.24	1.23
	0.750	1.52	1.52	1.51	1.49	1.48	1.47	1.46	1.45	1.44	1.43	1.41	1.40	1.38	1.36
	0.800	1.80	1.75	1.71	1.68	1.66	1.64	1.62	1.60	1.58	1.56	1.54	1.51	1.49	1.46
16	0.500	0.72	0.82	0.88	0.91	0.93	0.95	0.96	0.97	0.99	1.00	1.01	1.02	1.03	1.04
	0.600	0.97	1.04	1.08	1.10	1.11	1.12	1.12	1.13	1.13	1.14	1.14	1.14	1.14	1.14
	0.667	1.18	1.22	1.24	1.25	1.25	1.25	1.25	1.25	1.25	1.25	1.24	1.24	1.23	1.22
	0.750	1.51	1.51	1.50	1.48	1.47	1.46	1.45	1.44	1.43	1.41	1.40	1.38	1.37	1.34
	0.800	1.78	1.74	1.70	1.67	1.64	1.62	1.61	1.58	1.56	1.54	1.52	1.49	1.47	1.43
17	0.500	0.72	0.82	0.87	0.91	0.93	0.94	0.96	0.97	0.98	0.99	1.01	1.02	1.03	1.04
	0.600	0.97	1.04	1.07	1.09	1.10	1.11	1.12	1.12	1.13	1.13	1.13	1.14	1.14	1.14
	0.667	1.17	1.22	1.23	1.24	1.24	1.24	1.24	1.24	1.24	1.24	1.23	1.23	1.22	1.21
	0.750	1.51	1.50	1.49	1.47	1.46	1.45	1.44	1.43	1.41	1.40	1.39	1.37	1.36	1.33
	0.800	1.77	1.72	1.68	1.65	1.63	1.61	1.59	1.57	1.55	1.53	1.50	1.48	1.46	1.42

PERCENTAGE POINTS OF THE F DISTRIBUTION

$v_2 \backslash v_1$	q	2	3	4	5	6	7	8	10	12	15	20	30	50	∞
18	0.500	0.72	0.82	0.87	0.90	0.93	0.94	0.95	0.97	0.98	0.99	1.00	1.02	1.02	1.04
	0.600	0.96	1.04	1.07	1.09	1.10	1.11	1.11	1.12	1.12	1.13	1.13	1.13	1.13	1.13
	0.667	1.17	1.21	1.23	1.24	1.24	1.24	1.24	1.24	1.23	1.23	1.23	1.22	1.22	1.21
	0.750	1.50	1.49	1.48	1.46	1.45	1.44	1.43	1.42	1.40	1.39	1.38	1.36	1.34	1.32
	0.800	1.76	1.71	1.67	1.64	1.62	1.60	1.58	1.55	1.53	1.51	1.49	1.46	1.44	1.40
19	0.500	0.72	0.82	0.87	0.90	0.92	0.94	0.95	0.97	0.98	0.99	1.00	1.01	1.02	1.04
	0.600	0.96	1.03	1.07	1.09	1.10	1.10	1.11	1.12	1.12	1.12	1.13	1.13	1.13	1.13
	0.667	1.16	1.21	1.22	1.23	1.23	1.23	1.23	1.23	1.23	1.23	1.22	1.22	1.21	1.20
	0.750	1.49	1.49	1.47	1.46	1.44	1.43	1.42	1.41	1.40	1.38	1.37	1.35	1.33	1.30
	0.800	1.75	1.70	1.66	1.63	1.61	1.58	1.57	1.54	1.52	1.50	1.48	1.45	1.43	1.39
20	0.500	0.72	0.82	0.87	0.90	0.92	0.94	0.95	0.97	0.98	0.99	1.00	1.01	1.02	1.03
	0.600	0.96	1.03	1.06	1.08	1.09	1.10	1.11	1.11	1.12	1.12	1.12	1.12	1.12	1.12
	0.667	1.16	1.21	1.22	1.23	1.23	1.23	1.23	1.23	1.22	1.22	1.22	1.21	1.20	1.19
	0.750	1.49	1.48	1.47	1.45	1.44	1.43	1.42	1.40	1.39	1.37	1.36	1.34	1.32	1.29
	0.800	1.75	1.70	1.65	1.62	1.60	1.58	1.56	1.53	1.51	1.49	1.47	1.44	1.41	1.37
21	0.500	0.72	0.81	0.87	0.90	0.92	0.94	0.95	0.96	0.98	0.99	1.00	1.01	1.02	1.03
	0.600	0.96	1.03	1.06	1.08	1.09	1.10	1.10	1.11	1.11	1.12	1.12	1.12	1.12	1.12
	0.667	1.16	1.20	1.22	1.22	1.22	1.22	1.22	1.22	1.22	1.22	1.21	1.20	1.20	1.19
	0.750	1.48	1.48	1.46	1.44	1.43	1.42	1.41	1.39	1.38	1.37	1.35	1.33	1.32	1.28
	0.800	1.74	1.69	1.65	1.61	1.59	1.57	1.55	1.52	1.50	1.48	1.46	1.43	1.40	1.36
22	0.500	0.72	0.81	0.87	0.90	0.92	0.93	0.95	0.96	0.97	0.99	1.00	1.01	1.02	1.03
	0.600	0.96	1.03	1.06	1.08	1.09	1.10	1.10	1.11	1.11	1.11	1.12	1.12	1.12	1.12
	0.667	1.16	1.20	1.21	1.22	1.22	1.22	1.22	1.22	1.21	1.21	1.21	1.20	1.19	1.18
	0.750	1.48	1.47	1.45	1.44	1.42	1.41	1.40	1.39	1.37	1.36	1.34	1.32	1.31	1.28
	0.800	1.73	1.68	1.64	1.61	1.58	1.56	1.54	1.51	1.49	1.47	1.45	1.42	1.39	1.35
23	0.500	0.71	0.81	0.86	0.90	0.92	0.93	0.95	0.96	0.97	0.98	1.00	1.01	1.02	1.03
	0.600	0.95	1.02	1.06	1.07	1.09	1.09	1.10	1.10	1.11	1.11	1.11	1.11	1.11	1.11
	0.667	1.15	1.20	1.21	1.22	1.22	1.22	1.22	1.21	1.21	1.21	1.20	1.19	1.19	1.17
	0.750	1.47	1.47	1.45	1.43	1.42	1.41	1.40	1.38	1.37	1.35	1.34	1.32	1.30	1.27
	0.800	1.73	1.68	1.63	1.60	1.57	1.55	1.53	1.51	1.49	1.46	1.44	1.41	1.38	1.34
24	0.500	0.71	0.81	0.86	0.90	0.92	0.93	0.94	0.96	0.97	0.98	0.99	1.01	1.01	1.03
	0.600	0.95	1.02	1.06	1.07	1.08	1.09	1.10	1.10	1.10	1.11	1.11	1.11	1.11	1.11
	0.667	1.15	1.19	1.21	1.21	1.21	1.21	1.21	1.21	1.21	1.20	1.20	1.19	1.18	1.17
	0.750	1.47	1.46	1.44	1.43	1.41	1.40	1.39	1.38	1.36	1.35	1.33	1.31	1.29	1.26
	0.800	1.72	1.67	1.63	1.59	1.57	1.55	1.53	1.50	1.48	1.46	1.43	1.40	1.38	1.33
25	0.500	0.71	0.81	0.86	0.89	0.92	0.93	0.94	0.96	0.97	0.98	0.99	1.00	1.01	1.03
	0.600	0.95	1.02	1.05	1.07	1.08	1.09	1.09	1.10	1.10	1.11	1.11	1.11	1.11	1.11
	0.667	1.15	1.19	1.21	1.21	1.21	1.21	1.21	1.21	1.20	1.20	1.19	1.19	1.18	1.16
	0.750	1.47	1.46	1.44	1.42	1.41	1.40	1.39	1.37	1.36	1.34	1.33	1.31	1.29	1.25
	0.800	1.72	1.66	1.62	1.59	1.56	1.54	1.52	1.49	1.47	1.45	1.42	1.39	1.37	1.32
26	0.500	0.71	0.81	0.86	0.89	0.91	0.93	0.94	0.96	0.97	0.98	0.99	1.00	1.01	1.03
	0.600	0.95	1.02	1.05	1.07	1.08	1.09	1.09	1.10	1.10	1.10	1.10	1.11	1.11	1.10
	0.667	1.15	1.19	1.20	1.21	1.21	1.21	1.21	1.20	1.20	1.20	1.19	1.18	1.18	1.16
	0.750	1.46	1.45	1.44	1.42	1.41	1.39	1.38	1.37	1.35	1.34	1.32	1.30	1.28	1.25
	0.800	1.71	1.66	1.62	1.58	1.56	1.53	1.52	1.49	1.47	1.44	1.42	1.39	1.36	1.31

APPENDIX A. APPENDIX

PERCENTAGE POINTS OF THE F DISTRIBUTION

$v_2\backslash v_1$	q	2	3	4	5	6	7	8	10	12	15	20	30	50	∞
27	0.500	0.71	0.81	0.86	0.89	0.91	0.93	0.94	0.96	0.97	0.ga	0.99	1.00	1.01	1.03
	0.600	0.95	1.02	1.05	1.07	1.08	1.08	1.09	1.10	1.10	1.10	1.10	1.10	1.10	1.10
	0.667	1.14	1.19	1.20	1.21	1.21	1.21	1.20	1.20	1.20	1.19	1.19	1.18	1.17	1.16
	0.750	1.46	1.45	1.43	1.42	1.40	1.39	1.38	1.36	1.35	1.33	1.32	1.30	1.28	1.24
	0.800	1.71	1.66	1.61	1.58	1.55	1.53	1.51	1.48	1.46	1.44	1.41	1.3a	1.35	1.30
28	0.500	0.71	0.81	0.86	0.89	0.91	0.93	0.94	0.96	0.97	0.98	0.99	1.00	1.01	1.02
	0.600	0.95	1.02	1.05	1.07	1.08	1.08	1.09	1.09	1.10	1.10	1.10	1.10	1.10	1.10
	0.667	1.14	1.18	1.20	1.20	1.20	1.20	1.20	1.20	1.20	1.19	1.19	1.18	1.17	1.15
	0.750	1.46	1.45	1.43	1.41	1.40	1.39	1.38	1.36	1.34	1.33	1.31	1.29	1.27	1.24
	0.800	1.71	1.65	1.61	1.57	1.55	1.52	1.51	1.48	1.46	1.43	1.41	1.37	1.35	1.30
29	0.500	0.71	0.81	0.86	0.89	0.91	0.93	0.94	0.96	0.97	0.98	0.99	1.00	1.01	1.02
	0.600	0.95	1.02	1.05	1.06	1.08	1.08	1.09	1.09	1.10	1.10	1.10	1.10	1.10	1.10
	0.667	1.14	1.18	1.20	1.20	1.20	1.20	1.20	1.20	1.19	1.19	1.18	1.17	1.17	1.15
	0.750	1.45	1.45	1.43	1.41	1.40	1.38	1.37	1.35	1.34	1.32	1.31	1.29	1.27	1.23
	0.800	1.70	1.65	1.60	1.57	1.54	1.52	1.50	1.47	1.45	1.43	1.40	1.37	1.34	1.29
30	0.500	0.71	0.al	0.86	0.89	0.91	0.93	0.94	0.96	0.97	0.98	0.99	1.00	1.01	1.02
	0.600	0.94	1.01	1.05	1.06	1.07	1.08	1.08	1.09	1.09	1.10	1.10	1.10	1.10	1.09
	0.667	1.14	1.18	1.19	1.20	1.20	1.20	1.20	1.19	1.19	1.19	1.18	1.17	1.16	1.15
	0.750	1.45	1.44	1.42	1.41	1.39	1.38	1.37	1.35	1.34	1.32	1.30	1.28	1.26	1.23
	0.800	1.70	1.64	1.60	1.57	1.54	1.52	1.50	1.47	1.45	1.42	1.39	1.36	1.34	1.28
60	0.500	0.70	0.80	0.85	0.88	0.90	0.92	0.93	0.94	0.96	0.97	0.98	0.99	1.00	1.01
	0.600	0.93	1.00	1.03	1.04	1.05	1.06	1.06	1.07	1.07	1.07	1.07	1.07	1.07	1.06
	0.667	1.12	1.16	1.17	1.17	1.17	1.17	1.17	1.16	1.16	1.15	1.14	1.13	1.12	1.10
	0.750	1.42	1.41	1.38	1.37	1.35	1.33	1.32	1.30	1.29	1.27	1.25	1.22	1.20	1.15
	0.800	1.65	1.59	1.55	1.51	1.48	1.46	1.44	1.41	1.38	1.35	1.32	1.29	1.25	1.18
80	0.500	0.70	0.80	0.85	0.88	0.90	0.91	0.93	0.94	0.95	0.96	0.97	0.99	1.00	1.01
	0.600	0.93	0.99	1.02	1.04	1.05	1.06	1.06	1.06	1.07	1.07	1.07	1.06	1.06	1.05
	0.667	1.11	1.15	1.16	1.17	1.17	1.16	1.16	1.16	1.15	1.14	1.13	1.12	1.11	1.08
	0.750	1.41	1.40	1.38	1.36	1.34	1.32	1.31	1.29	1.27	1.26	1.23	1.21	1.18	1.12
	0.800	1.64	1.58	1.53	1.50	1.47	1.44	1.42	1.39	1.37	1.34	1.31	1.27	1.23	1.16
100	0.500	0.70	0.79	0.84	0.88	0.90	0.91	0.92	0.94	0.95	0.96	0.97	0.98	0.99	1.01
	0.600	0.92	0.99	1.02	1.04	1.05	1.05	1.06	1.06	1.06	1.06	1.06	1.06	1.06	1.04
	0.667	1.11	1.15	1.16	1.16	1.16	1.16	1.16	1.15	1.15	1.14	1.13	1.12	1.10	1.07
	0.750	1.41	1.39	1.37	1.35	1.33	1.32	1.30	1.28	1.27	1.25	1.23	1.20	1.17	1.11
	0.800	1.64	1.58	1.53	1.49	1.46	1.43	1.41	1.38	1.36	1.33	1.30	1.26	1.22	1.14
120	0.500	0.70	0.79	0.84	0.88	0.90	0.91	0.92	0.94	0.95	0.96	0.97	0.98	0.99	1.01
	0.600	0.92	0.99	1.02	1.04	1.04	1.05	1.05	1.06	1.06	1.06	1.06	1.06	1.05	1.04
	0.667	1.11	1.15	1.16	1.16	1.16	1.16	1.15	1.15	1.14	1.13	1.13	1.11	1.10	1.06
	0.750	1.40	1.39	1.37	1.35	1.33	1.31	1.30	1.28	1.26	1.24	1.22	1.19	1.16	1.10
	0.800	1.63	1.57	1.52	1.48	1.45	1.43	1.41	1.37	1.35	1.32	1.29	1.25	1.21	1.12
∞	0.500	0.69	0.79	0.84	0.87	0.89	0.91	0.92	0.93	0.95	0.96	0.97	0.98	0.99	1.00
	0.600	0.92	0.98	1.01	1.03	1.04	1.04	1.04	1.05	1.05	1.05	1.05	1.04	1.04	1.00
	0.667	1.10	1.13	1.14	1.15	1.14	1.14	1.14	1.13	1.13	1.12	1.11	1.09	1.07	1.00
	0.750	1.39	1.37	1.35	1.33	1.31	1.29	1.28	1.25	1.24	1.22	1.19	1.16	1.13	1.00
	0.800	1.61	1.55	1.50	1.46	1.43	1.40	1.38	1.34	1.32	1.29	1.25	1.21	1.16	1.00

PERCENTAGE POINTS OF THE F DISTRIBUTION

$v_2 \backslash v_1$	q	2	3	4	5	6	7	8	10	12	15	20	30	50	∞
1	0.900	49.5	53.6	55.8	57.2	58.2	59.1	59.7	60.5	61.0	61.5	62.0	62.6	63.0	63.3
	0.950	199.	216.	225.	230.	234.	237.	239.	242.	244.	246.	248.	250.	252.	254.
	0.975	800.	864.	900.	922.	937.	948.	957.	969.	977.	985.	993.			
	0.990														
	0.999														
2	0.900	9.00	9.16	9.24	9.29	9.33	9.35	9.37	9.39	9.41	9.43	9.44	9.46	9.47	9.49
	0.950	19.0	19.2	19.2	19.3	19.3	19.4	19.4	19.4	19.4	19.4	19.4	19.5	19.5	19.5
	0.975	39.0	39.2	39.2	39.3	39.3	39.4	39.4	39.4	39.4	39.4	39.4	39.5	39.5	39.5
	0.990	99.0	99.2	99.2	99.3	99.3	99.4	100.	100.	100.	100.	100.	100.	100.	99.5
	0.999	999.	999.												
3	0.900	5.46	5.39	5.34	5.31	5.28	5.27	5.25	5.23	5.22	5.20	5.18	5.17	5.15	5.13
	0.950	9.55	9.28	9.12	9.01	8.94	8.89	8.85	8.79	8.74	8.70	8.66	8.62	8.58	8.53
	0.975	16.0	15.4	15.1	14.9	14.7	14.6	14.5	14.4	14.3	14.3	14.2	14.1	14.0	13.9
	0.990	30.8	29.5	28.7	28.2	27.9	27.7	27.5	27.2	27.1	26.9	26.7	26.5	26.4	26.1
	0.999	149.	141.	137.	135.	133.	132.	131.	129.	128.	127.	126.	125.	125.	123.
4	0.900	4.32	4.19	4.11	4.05	4.01	3.98	3.95	3.92	3.90	3.87	3.84	3.82	3.79	3.76
	0.950	6.94	6.59	6.39	6.26	6.16	6.09	6.04	5.96	5.91	5.86	5.80	5.75	5.70	5.63
	0.975	10.6	9.98	9.60	9.36	9.20	9.07	8.98	8.84	8.75	8.66	8.56	8.46	8.38	8.26
	0.990	18.0	16.7	16.0	15.5	15.2	15.0	14.8	14.5	14.4	14.2	14.0	13.8	13.7	13.5
	0.999	61.2	56.2	53.4	51.7	50.5	49.7	49.0	48.0	47.4	46.8	46.1	45.4	44.9	44.1
5	0.900	3.78	3.62	3.52	3.45	3.40	3.37	3.34	3.30	3.27	3.24	3.21	3.17	3.15	3.10
	0.950	5.79	5.41	5.19	5.05	4.95	4.88	4.82	4.74	4.68	4.62	4.56	4.50	4.44	4.36
	0.975	8.43	7.76	7.39	7.15	6.98	6.85	6.76	6.62	6.52	6.43	6.33	6.23	6.14	6.02
	0.990	13.3	12.1	11.4	11.0	10.7	10.5	10.3	10.1	9.89	9.72	9.55	9.38	9.24	9.02
	0.999	37.1	33.2	31.1	29.8	28.8	28.2	27.6	26.9	26.4	25.9	25.4	24.9	24.4	23.8
6	0.900	3.46	3.29	3.18	3.11	3.05	3.01	2.98	2.94	2.90	2.87	2.84	2.80	2.77	2.72
	0.950	5.14	4.76	4.53	4.39	4.28	4.21	4.15	4.06	4.00	3.94	3.87	3.81	3.75	3.67
	0.975	7.26	6.60	6.23	5.99	5.82	5.70	5.60	5.46	5.37	5.27	5.17	5.07	4.98	4.85
	0.990	10.9	9.78	9.15	8.75	8.47	8.26	8.10	7.87	7.72	7.56	7.40	7.23	7.09	6.88
	0.999	27.0	23.7	21.9	20.8	20.0	19.5	19.0	18.4	18.0	17.6	17.1	16.7	16.3	15.7
7	0.900	3.26	3.07	2.96	2.88	2.83	2.78	2.75	2.70	2.67	2.63	2.59	2.56	2.52	2.47
	0.950	4.74	4.35	4.12	3.97	3.87	3.79	3.73	3.64	3.57	3.51	3.44	3.38	3.32	3.23
	0.975	6.54	5.89	5.52	5.29	5.12	4.99	4.90	4.76	4.67	4.57	4.47	4.36	4.28	4.14
	0.990	9.55	8.45	7.85	7.46	7.19	6.99	6.84	6.62	6.47	6.31	6.16	5.99	5.86	5.65
	0.999	21.7	18.8	17.2	16.2	15.5	15.0	14.6	14.1	13.7	13.3	12.9	12.5	12.2	11.7
8	0.900	3.11	2.92	2.81	2.73	2.67	2.62	2.59	2.54	2.50	2.46	2.42	2.38	2.35	2.29
	0.950	4.46	4.07	3.84	3.69	3.58	3.50	3.44	3.35	3.28	3.22	3.15	3.08	3.02	2.93
	0.975	6.06	5.42	5.05	4.82	4.65	4.53	4.43	4.29	4.20	4.10	4.00	3.89	3.81	3.67
	0.990	8.65	7.59	7.01	6.63	6.37	6.18	6.03	5.81	5.67	5.52	5.36	5.20	5.07	4.86
	0.999	18.5	15.8	14.4	13.5	12.9	12.4	12.0	11.5	11.2	10.8	10.5	10.1	9.80	9.33

APPENDIX A. APPENDIX

PERCENTAGE POINTS OF THE F DISTRIBUTION

$v_2\backslash v_1$	q	2	3	4	5	6	7	8	10	12	15	20	30	50	∞
9	0.900	3.01	2.81	2.69	2.61	2.55	2.51	2.47	2.42	2.38	2.34	2.30	2.25	2.22	2.16
	0.950	4.26	3.86	3.63	3.48	3.37	3.29	3.23	3.14	3.07	3.01	2.94	2.86	2.80	2.71
	0.975	5.71	5.08	4.72	4.48	4.32	4.20	4.10	3.96	3.87	3.77	3.67	3.56	3.47	3.33
	0.990	8.02	6.99	6.42	6.06	5.80	5.61	5.47	5.26	5.11	4.96	4.81	4.65	4.52	4.31
	0.999	16.4	13.9	12.6	11.7	11.1	10.7	10.4	9.89	9.57	9.24	8.90	8.55	8.26	7.81
10	0.900	2.92	2.73	2.61	2.52	2.46	2.41	2.38	2.32	2.28	2.24	2.20	2.16	2.12	2.06
	0.950	4.10	3.71	3.48	3.33	3.22	3.14	3.07	2.98	2.91	2.84	2.77	2.70	2.64	2.54
	0.975	5.46	4.83	4.47	4.24	4.07	3.95	3.85	3.72	3.62	3.52	3.42	3.31	3.22	3.08
	0.990	7.56	6.55	5.99	5.64	5.39	5.20	5.06	4.85	4.71	4.56	4.41	4.25	4.11	3.91
	0.999	14.9	12.6	11.3	10.5	9.93	9.52	9.20	8.75	8.45	8.13	7.80	7.47	7.19	6.76
11	0.900	2.86	2.66	2.54	2.45	2.39	2.34	2.30	2.25	2.21	2.17	2.12	2.08	2.04	1.97
	0.950	3.98	3.59	3.36	3.20	3.09	3.01	2.95	2.85	2.79	2.72	2.65	2.57	2.51	2.40
	0.975	5.26	4.63	4.28	4.04	3.88	3.76	3.66	3.53	3.43	3.33	3.23	3.12	3.03	2.88
	0.990	7.21	6.22	5.67	5.32	5.07	4.89	4.74	4.54	4.40	4.25	4.10	3.94	3.81	3.60
	0.999	13.8	11.6	10.3	9.58	9.05	8.66	8.35	7.92	7.63	7.32	7.01	6.68	6.42	6.00
12	0.900	2.81	2.61	2.48	2.39	2.33	2.28	2.24	2.19	2.15	2.10	2.06	2.01	1.97	1.90
	0.950	3.89	3.49	3.26	3.11	3.00	2.91	2.85	2.75	2.69	2.62	2.54	2.47	2.40	2.30
	0.975	5.10	4.47	4.12	3.89	3.73	3.61	3.51	3.37	3.28	3.18	3.07	2.96	2.87	2.72
	0.990	6.93	5.95	5.41	5.06	4.82	4.64	4.50	4.30	4.16	4.01	3.86	3.70	3.57	3.36
	0.999	13.0	10.8	9.63	8.89	8.38	8.00	7.71	7.29	7.00	6.71	6.40	6.09	5.83	5.42
13	0.900	2.76	2.56	2.43	2.35	2.28	2.23	2.20	2.14	2.10	2.05	2.01	1.96	1.92	1.85
	0.950	3.81	3.41	3.18	3.03	2.92	2.83	2.77	2.67	2.60	2.53	2.46	2.38	2.31	2.21
	0.975	4.97	4.35	4.00	3.77	3.60	3.48	3.39	3.25	3.15	3.05	2.95	2.84	2.74	2.60
	0.990	6.70	5.74	5.21	4.86	4.62	4.44	4.30	4.10	3.96	3.82	3.66	3.51	3.37	3.17
	0.999	12.3	10.2	9.07	8.35	7.86	7.49	7.21	6.80	6.52	6.23	5.93	5.63	5.37	4.97
14	0.900	2.73	2.52	2.39	2.31	2.24	2.19	2.15	2.10	2.05	2.01	1.96	1.91	1.87	1.80
	0.950	3.74	3.34	3.11	2.96	2.85	2.76	2.70	2.60	2.53	2.46	2.39	2.31	2.24	2.13
	0.975	4.86	4.24	3.89	3.66	3.50	3.38	3.29	3.15	3.05	2.95	2.84	2.73	2.64	2.49
	0.990	6.51	5.56	5.04	4.69	4.46	4.28	4.14	3.94	3.80	3.66	3.51	3.35	3.22	3.00
	0.999	11.8	9.73	8.62	7.92	7.44	7.08	6.80	6.40	6.13	5.85	5.56	5.25	5.00	4.60
15	0.900	2.70	2.49	2.36	2.27	2.21	2.16	2.12	2.06	2.02	1.97	1.92	1.87	1.83	1.76
	0.950	3.68	3.29	3.06	2.90	2.79	2.71	2.64	2.54	2.48	2.40	2.33	2.25	2.18	2.07
	0.975	4.77	4.15	3.80	3.58	3.41	3.29	3.20	3.06	2.96	2.86	2.76	2.64	2.55	2.40
	0.990	6.36	5.42	4.89	4.56	4.32	4.14	4.00	3.80	3.67	3.52	3.37	3.21	3.08	2.87
	0.999	11.3	9.34	8.25	7.57	7.09	6.74	6.47	6.08	5.81	5.53	5.25	4.95	4.70	4.31
16	0.900	2.67	2.46	2.33	2.24	2.18	2.13	2.09	2.03	1.99	1.94	1.89	1.84	1.79	1.72
	0.950	3.63	3.24	3.01	2.85	2.74	2.66	2.59	2.49	2.42	2.35	2.28	2.19	2.12	2.01
	0.975	4.69	4.08	3.73	3.50	3.34	3.22	3.12	2.99	2.89	2.79	2.68	2.57	2.47	2.32
	0.990	6.23	5.29	4.77	4.44	4.20	4.03	3.89	3.69	3.55	3.41	3.26	3.10	2.97	2.75
	0.999	11.0	9.01	7.94	7.27	6.80	6.46	6.19	5.81	5.55	5.27	4.99	4.70	4.45	4.06
17	0.900	2.64	2.44	2.31	2.22	2.15	2.10	2.06	2.00	1.96	1.91	1.86	1.81	1.76	1.69
	0.950	3.59	3.20	2.96	2.81	2.70	2.61	2.55	2.45	2.38	2.31	2.23	2.15	2.08	1.96
	0.975	4.62	4.01	3.66	3.44	3.28	3.16	3.06	2.92	2.82	2.72	2.62	2.50	2.41	2.25
	0.990	6.11	5.18	4.67	4.34	4.10	3.93	3.79	3.59	3.46	3.31	3.16	3.00	2.87	2.65
	0.999	10.7	8.73	7.68	7.02	6.56	6.22	5.96	5.58	5.32	5.05	4.77	4.48	4.24	3.85

PERCENTAGE POINTS OF THE F DISTRIBUTION

$v_2 \backslash v_1$	q	2	3	4	5	6	7	8	10	12	15	20	30	50	∞
18	0.900	2.62	2.42	2.29	2.20	2.13	2.08	2.04	1.98	1.93	1.89	1.84	1.78	1.74	1.66
	0.950	3.55	3.16	2.93	2.77	2.66	2.58	2.51	2.41	2.34	2.27	2.19	2.11	2.04	1.92
	0.975	4.56	3.95	3.61	3.38	3.22	3.10	3.01	2.87	2.77	2.67	2.56	2.44	2.35	2.19
	0.990	6.01	5.09	4.58	4.25	4.01	3.84	3.71	3.51	3.37	3.23	3.08	2.92	2.78	2.57
	0.999	10.4	8.49	7.46	6.81	6.35	6.02	5.76	5.39	5.13	4.87	4.59	4.30	4.06	3.67
19	0.900	2.61	2.40	2.27	2.18	2.11	2.06	2.02	1.96	1.91	1.86	1.81	1.76	1.71	1.63
	0.950	3.52	3.13	2.90	2.74	2.63	2.54	2.48	2.38	2.31	2.23	2.16	2.07	2.00	1.88
	0.975	4.51	3.90	3.56	3.33	3.17	3.05	2.96	2.82	2.72	2.62	2.51	2.39	2.30	2.13
	0.990	5.93	5.01	4.50	4.17	3.94	3.77	3.63	3.43	3.30	3.15	3.00	2.84	2.71	2.49
	0.999	10.2	8.28	7.27	6.62	6.18	5.85	5.59	5.22	4.97	4.70	4.43	4.14	3.90	3.51
20	0.900	2.59	2.38	2.25	2.16	2.09	2.04	2.00	1.94	1.89	1.84	1.79	1.74	1.69	1.61
	0.950	3.49	3.10	2.87	2.71	2.60	2.51	2.45	2.35	2.28	2.20	2.12	2.04	1.97	1.84
	0.975	4.46	3.86	3.51	3.29	3.13	3.01	2.91	2.77	2.68	2.57	2.46	2.35	2.25	2.09
	0.990	5.85	4.94	4.43	4.10	3.87	3.70	3.56	3.37	3.23	3.09	2.94	2.78	2.64	2.42
	0.999	9.95	8.10	7.10	6.46	6.02	5.69	5.44	5.08	4.82	4.56	4.29	4.00	3.76	3.38
21	0.900	2.57	2.36	2.23	2.14	2.08	2.02	1.98	1.92	1.87	1.83	1.78	1.72	1.67	1.59
	0.950	3.47	3.07	2.84	2.68	2.57	2.49	2.42	2.32	2.25	2.18	2.10	2.01	1.94	1.81
	0.975	4.42	3.82	3.48	3.25	3.09	2.97	2.87	2.73	2.64	2.53	2.42	2.31	2.21	2.04
	0.990	5.78	4.87	4.37	4.04	3.81	3.64	3.51	3.31	3.17	3.03	2.88	2.72	2.58	2.36
	0.999	9.77	7.94	6.95	6.32	5.88	5.56	5.31	4.95	4.70	4.44	4.17	3.88	3.64	3.26
22	0.900	2.56	2.35	2.22	2.13	2.06	2.01	1.97	1.90	1.86	1.81	1.76	1.70	1.65	1.57
	0.950	3.44	3.05	2.82	2.66	2.55	2.46	2.40	2.30	2.23	2.15	2.07	1.98	1.91	1.78
	0.975	4.38	3.78	3.44	3.22	3.05	2.93	2.84	2.70	2.60	2.50	2.39	2.27	2.17	2.00
	0.990	5.72	4.82	4.31	3.99	3.76	3.59	3.45	3.26	3.12	2.98	2.83	2.67	2.53	2.31
	0.999	9.61	7.80	6.81	6.19	5.76	5.44	5.19	4.83	4.58	4.33	4.06	3.78	3.54	3.15
23	0.900	2.55	2.34	2.21	2.11	2.05	1.99	1.95	1.89	1.84	1.80	1.74	1.69	1.64	1.55
	0.950	3.42	3.03	2.80	2.64	2.53	2.44	2.37	2.27	2.20	2.13	2.05	1.96	1.88	1.76
	0.975	4.35	3.75	3.41	3.18	3.02	2.90	2.81	2.67	2.57	2.47	2.36	2.24	2.14	1.97
	0.990	5.66	4.76	4.26	3.94	3.71	3.54	3.41	3.21	3.07	2.93	2.78	2.62	2.48	2.26
	0.999	9.47	7.67	6.70	6.08	5.65	5.33	5.09	4.73	4.48	4.23	3.96	3.68	3.44	3.05
24	0.900	2.54	2.33	2.19	2.10	2.04	1.98	1.94	1.88	1.83	1.78	1.73	1.67	1.62	1.53
	0.950	3.40	3.01	2.78	2.62	2.51	2.42	2.36	2.25	2.18	2.11	2.03	1.94	1.86	1.73
	0.975	4.32	3.72	3.38	3.15	2.99	2.87	2.78	2.64	2.54	2.44	2.33	2.21	2.11	1.94
	0.990	5.61	4.72	4.22	3.90	3.67	3.50	3.36	3.17	3.03	2.89	2.74	2.58	2.44	2.21
	0.999	9.34	7.55	6.59	5.98	5.55	5.23	4.99	4.64	4.39	4.14	3.87	3.59	3.36	2.97
25	0.900	2.53	2.32	2.18	2.09	2.02	1.97	1.93	1.87	1.82	1.77	1.72	1.66	1.61	1.52
	0.950	3.39	2.99	2.76	2.60	2.49	2.40	2.34	2.24	2.16	2.09	2.01	1.92	1.84	1.71
	0.975	4.29	3.69	3.35	3.13	2.97	2.85	2.75	2.61	2.51	2.41	2.30	2.18	2.08	1.91
	0.990	5.57	4.68	4.18	3.85	3.63	3.46	3.32	3.13	2.99	2.85	2.70	2.54	2.40	2.17
	0.999	9.22	7.45	6.49	5.89	5.46	5.15	4.91	4.56	4.31	4.06	3.79	3.52	3.28	2.89
26	0.900	2.52	2.31	2.17	2.08	2.01	1.96	1.92	1.86	1.81	1.76	1.71	1.65	1.59	1.50
	0.950	3.37	2.98	2.74	2.59	2.47	2.39	2.32	2.22	2.15	2.07	1.99	1.90	1.82	1.69184
	0.975	4.27	3.67	3.33	3.10	2.94	2.82	2.73	2.59	2.49	2.39	2.28	2.16	2.05	1.88
	0.990	5.53	4.64	4.14	3.82	3.59	3.42	3.29	3.09	2.96	2.81	2.66	2.50	2.36	2.13
	0.999	9.12	7.36	6.41	5.80	5.38	5.07	4.83	4.48	4.24	3.99	3.72	3.44	3.21	2.82

APPENDIX A. APPENDIX

PERCENTAGE POINTS OF THE F DISTRIBUTION

$v_2 \backslash v_1$	q	2	3	4	5	6	7	8	10	12	15	20	30	50	∞
27	0.900	2.51	2.30	2.17	2.07	2.00	1.95	1.91	1.85	1.80	1.75	1.70	1.64	1.58	1.49
	0.950	3.35	2.96	2.73	2.57	2.46	2.37	2.31	2.20	2.13	2.06	1.97	1.88	1.81	1.67
	0.975	4.24	3.65	3.31	3.08	2.92	2.80	2.71	2.57	2.47	2.36	2.25	2.13	2.03	1.85
	0.990	5.49	4.60	4.11	3.78	3.56	3.39	3.26	3.06	2.93	2.78	2.63	2.47	2.33	2.10
	0.999	9.02	7.27	6.33	5.73	5.31	5.00	4.76	4.41	4.17	3.92	3.66	3.38	3.14	2.75
28	0.900	2.50	2.29	2.16	2.06	2.00	1.94	1.90	1.84	1.79	1.74	1.69	1.63	1.57	1.48
	0.950	3.34	2.95	2.71	2.56	2.45	2.36	2.29	2.19	2.12	2.04	1.96	1.87	1.79	1.65
	0.975	4.22	3.63	3.29	3.06	2.90	2.78	2.69	2.55	2.45	2.34	2.23	2.11	2.01	1.83
	0.990	5.45	4.57	4.07	3.75	3.53	3.36	3.23	3.03	2.90	2.75	2.60	2.44	2.30	2.06
	0.999	8.93	7.19	6.25	5.66	5.24	4.93	4.69	4.35	4.11	3.86	3.60	3.32	3.09	2.69
29	0.900	2.50	2.28	2.15	2.06	1.99	1.93	1.89	1.83	1.78	1.73	1.68	1.62	1.56	1.47
	0.950	3.33	2.93	2.70	2.55	2.43	2.35	2.28	2.18	2.10	2.03	1.94	1.85	1.77	1.64
	0.975	4.20	3.61	3.27	3.04	2.88	2.76	2.67	2.53	2.43	2.32	2.21	2.09	1.99	1.81
	0.990	5.42	4.54	4.04	3.73	3.50	3.33	3.20	3.00	2.87	2.73	2.57	2.41	2.27	2.03
	0.999	8.85	7.12	6.19	5.59	5.18	4.87	4.64	4.29	4.05	3.80	3.54	3.27	3.03	2.64
30	0.900	2.49	2.28	2.14	2.05	1.98	1.93	1.88	1.82	1.77	1.72	1.67	1.61	1.55	1.46
	0.950	3.32	2.92	2.69	2.53	2.42	2.33	2.27	2.16	2.09	2.01	1.93	1.84	1.76	1.62
	0.975	4.18	3.59	3.25	3.03	2.87	2.75	2.65	2.51	2.41	2.31	2.20	2.07	1.97	1.79
	0.990	5.39	4.51	4.02	3.70	3.47	3.30	3.17	2.98	2.84	2.70	2.55	2.39	2.25	2.01
	0.999	8.77	7.05	6.12	5.53	5.12	4.82	4.58	4.24	4.00	3.75	3.49	3.22	2.98	2.59
60	0.900	2.39	2.18	2.04	1.95	1.87	1.82	1.77	1.71	1.66	1.60	1.54	1.48	1.41	1.29
	0.950	3.15	2.76	2.53	2.37	2.25	2.17	2.10	1.99	1.92	1.84	1.75	1.65	1.56	1.39
	0.975	3.93	3.34	3.01	2.79	2.63	2.51	2.41	2.27	2.17	2.06	1.94	1.82	1.70	1.48
	0.990	4.98	4.13	3.65	3.34	3.12	2.95	2.82	2.63	2.50	2.35	2.20	2.03	1.88	1.60
	0.999	7.77	6.17	5.31	4.76	4.37	4.09	3.86	3.54	3.32	3.08	2.83	2.55	2.32	1.89
80	0.900	2.37	2.15	2.02	1.92	1.85	1.79	1.75	1.68	1.63	1.57	1.51	1.44	1.38	1.24
	0.950	3.11	2.72	2.49	2.33	2.21	2.13	2.06	1.95	1.88	1.79	1.70	1.60	1.51	1.32
	0.975	3.86	3.28	2.95	2.73	2.57	2.45	2.35	2.21	2.11	2.00	1.88	1.75	1.63	1.40
	0.990	4.88	4.04	3.56	3.26	3.04	2.87	2.74	2.55	2.42	2.27	2.12	1.94	1.79	1.49
	0.999	7.54	5.97	5.12	4.58	4.20	3.92	3.70	3.39	3.16	2.93	2.68	2.41	2.16	1.72
100	0.900	2.36	2.14	2.00	1.91	1.83	1.78	1.73	1.66	1.61	1.56	1.49	1.42	1.35	1.21
	0.950	3.09	2.70	2.46	2.31	2.19	2.10	2.03	1.93	1.85	1.77	1.68	1.57	1.48	1.28
	0.975	3.83	3.25	2.92	2.70	2.54	2.42	2.32	2.18	2.08	1.97	1.85	1.71	1.59	1.35
	0.990	4.82	3.98	3.51	3.21	2.99	2.82	2.69	2.50	2.37	2.22	2.07	1.89	1.74	1.43
	0.999	7.41	5.86	5.02	4.48	4.11	3.83	3.61	3.30	3.07	2.84	2.59	2.32	2.08	1.62
120	0.900	2.35	2.13	1.99	1.90	1.82	1.77	1.72	1.65	1.60	1.54	1.48	1.41	1.34	1.19
	0.950	3.07	2.68	2.45	2.29	2.18	2.09	2.02	1.91	1.83	1.75	1.66	1.55	1.46	1.25
	0.975	3.80	3.23	2.89	2.67	2.52	2.39	2.30	2.16	2.05	1.94	1.82	1.69	1.56	1.31
	0.990	4.79	3.95	3.48	3.17	2.96	2.79	2.66	2.47	2.34	2.19	2.03	1.86	1.70	1.38
	0.999	7.32	5.78	4.95	4.42	4.04	3.77	3.55	3.24	3.02	2.78	2.53	2.26	2.02	1.54
∞	0.900	2.30	2.08	1.94	1.85	1.77	1.72	1.67	1.60	1.55	1.49	1.42	1.34	1.26	1.00
	0.950	3.00	2.60	2.37	2.21	2.10	2.01	1.94	1.83	1.75	1.67	1.57	1.46	1.35	1.00
	0.975	3.69	3.12	2.79	2.57	2.41	2.29	2.19	2.05	1.94	1.83	1.71	1.57	1.43	1.00
	0.990	4.61	3.78	3.32	3.02	2.80	2.64	2.51	2.32	2.18	2.04	1.88	1.70	1.52	1.00
	0.999	6.91	5.42	4.62	4.10	3.74	3.47	3.27	2.96	2.74	2.51	2.27	1.99	1.73	1.00

CHI-SQUARED PERCENTAGE POINTS

v	0.1%	0.5%	1.0%	2.5%	5.0%	10.0%	12.5%	20.0%	25.0%	33.3%	50.0%
1	0.000	0.000	0.000	0.001	0.004	0.016	0.025	0.064	0.102	0.186	0.455
2	0.002	0.010	0.020	0.051	0.103	0.211	0.267	0.446	0.575	0.811	1.386
3	0.024	0.072	0.115	0.216	0.352	0.584	0.692	1.005	1.213	1.568	2.366
4	0.091	0.207	0.297	0.484	0.711	1.064	1.219	1.649	1.923	2.378	3.357
5	0.210	0.412	0.554	0.831	1.145	1.610	1.808	2.343	2.675	3.216	4.351
6	0.381	0.676	0.872	1.237	1.635	2.204	2.441	3.070	3.455	4.074	5.348
7	0.598	0.989	1.239	1.690	2.167	2.833	3.106	3.822	4.255	4.945	6.346
8	0.857	1.344	1.646	2.180	2.733	3.490	3.797	4.594	5.071	5.826	7.344
9	1.152	1.735	2.088	2.700	3.325	4.168	4.507	5.380	5.899	6.716	8.343
10	1.479	2.156	2.558	3.247	3.940	4.865	5.234	6.179	6.737	7.612	9.342
11	1.834	2.603	3.053	3.816	4.575	5.578	5.975	6.989	7.584	8.514	10.341
12	2.214	3.074	3.571	4.404	5.226	6.304	6.729	7.807	8.438	9.420	11.340
13	2.617	3.565	4.107	5.009	5.892	7.042	7.493	8.634	9.299	10.331	12.340
14	3.041	4.075	4.660	5.629	6.571	7.790	8.266	9.467	10.165	11.245	13.339
15	3.483	4.601	5.229	6.262	7.261	8.547	9.048	10.307	11.037	12.163	14.339
16	3.942	5.142	5.812	6.908	7.962	9.312	9.837	11.152	11.912	13.083	15.338
17	4.416	5.697	6.408	7.564	8.672	10.085	10.633	12.002	12.792	14.006	16.338
18	4.905	6.265	7.015	8.231	9.390	10.865	11.435	12.857	13.675	14.931	17.338
19	5.407	6.844	7.633	8.907	10.117	11.651	12.242	13.716	14.562	15.859	18.338
20	5.921	7.434	8.260	9.591	10.851	12.443	13.055	14.578	15.452	16.788	19.337
21	6.447	8.034	8.897	10.283	11.591	13.240	13.873	15.445	16.344	17.720	20.337
22	6.983	8.643	9.542	10.982	12.338	14.041	14.695	16.314	17.240	18.653	21.337
23	7.529	9.260	10.196	11.689	13.091	14.848	15.521	17.187	18.137	19.587	22.337
24	8.085	9.886	10.856	12.401	13.848	15.659	16.351	18.062	19.037	20.523	23.337
25	8.649	10.520	11.524	13.120	14.611	16.473	17.184	18.940	19.939	21.461	24.337
26	9.222	11.160	12.198	13.844	15.379	17.292	18.021	19.820	20.843	22.399	25.336
27	9.803	11.808	12.879	14.573	16.151	18.114	18.861	20.703	21.749	23.339	26.336
28	10.391	12.461	13.565	15.308	16.928	18.939	19.704	21.588	22.657	24.280	27.336
29	10.986	13.121	14.256	16.047	17.708	19.768	20.550	22.475	23.567	25.222	28.336
30	11.588	13.787	14.953	16.791	18.493	20.599	21.399	23.364	24.478	26.165	29.336
35	14.688	17.192	18.509	20.569	22.465	24.797	25.678	27.836	29.054	30.894	34.336
40	17.916	20.707	22.164	24.433	26.509	29.051	30.008	32.345	33.660	35.643	39.335
45	21.251	24.311	25.901	28.366	30.612	33.350	34.379	36.884	38.291	40.407	44.335
50	24.674	27.991	29.707	32.357	34.764	37.689	38.785	41.449	42.942	45.184	49.335
55	28.173	31.735	33.570	36.398	38.958	42.060	43.220	46.036	47.610	49.972	54.335
60	31.738	35.534	37.485	40.482	43.188	46.459	47.680	50.641	52.294	54.770	59.335

APPENDIX A. APPENDIX

CHI-SQUARED PERCENTAGE POINTS

ν	60.0%	66.7%	75.0%	80.0%	87.5%	90.0%	95.0%	97.5%	99.0%	99.5%	99.9%
1	0.708	0.936	1.323	1.642	2.354	2.706	3.841	5.024	6.635	7.879	10.828
2	1.833	2.197	2.773	3.219	4.159	4.605	5.991	7.378	9.210	10.597	13.816
3	2.946	3.405	4.108	4.642	5.739	6.251	7.815	9.348	11.345	12.838	16.266
4	4.045	4.579	5.385	5.989	7.214	7.779	9.488	11.143	13.277	14.860	18.467
5	5.132	5.730	6.626	7.289	8.625	9.236	11.070	12.833	15.086	16.750	20.515
6	6.211	6.867	7.841	8.558	9.992	10.645	12.592	14.449	16.812	18.548	22.458
7	7.283	7.992	9.037	9.803	11.326	12.017	14.067	16.013	18.475	20.278	24.322
8	8.351	9.107	10.219	11.030	12.636	13.362	15.507	17.535	20.090	21.955	26.125
9	9.414	10.215	11.389	12.242	13.926	14.684	16.919	19.023	21.666	23.589	27.877
10	10.473	11.317	12.549	13.442	15.198	15.987	18.307	20.483	23.209	25.188	29.588
11	11.530	12.414	13.701	14.631	16.457	17.275	19.675	21.920	24.725	26.757	31.264
12	12.584	13.506	14.845	15.812	17.703	18.549	21.026	23.337	26.217	28.300	32.910
13	13.636	14.595	15.984	16.985	18.939	19.812	22.362	24.736	27.688	29.819	34.528
14	14.685	15.680	17.117	18.151	20.166	21.064	23.685	26.119	29.141	31.319	36.123
15	15.733	16.761	18.245	19.311	21.384	22.307	24.996	27.488	30.578	32.801	37.697
16	16.780	17.840	19.369	20.465	22.595	23.542	26.296	28.845	32.000	34.267	39.252
17	17.824	18.917	20.489	21.615	23.799	24.769	27.587	30.191	33.409	35.718	40.790
18	18.868	19.991	21.605	22.760	24.997	25.989	28.869	31.526	34.805	37.156	42.312
19	19.910	21.063	22.718	23.900	26.189	27.204	30.144	32.852	36.191	38.582	43.820
20	20.951	22.133	23.828	25.038	27.376	28.412	31.410	34.170	37.566	39.997	45.315
21	21.991	23.201	24.935	26.171	28.559	29.615	32.671	35.479	38.932	41.401	46.797
22	23.031	24.268	26.039	27.301	29.737	30.813	33.924	36.781	40.289	42.796	48.268
23	24.069	25.333	27.141	28.429	30.911	32.007	35.172	38.076	41.638	44.181	49.728
24	25.106	26.397	28.241	29.553	32.081	33.196	36.415	39.364	42.980	45.559	51.179
25	26.143	27.459	29.339	30.675	33.247	34.382	37.652	40.646	44.314	46.928	52.620
26	27.179	28.520	30.435	31.795	34.410	35.563	38.885	41.923	45.642	48.290	54.052
27	28.214	29.580	31.528	32.912	35.570	36.741	40.113	43.195	46.963	49.645	55.476
28	29.249	30.639	32.620	34.027	36.727	37.916	41.337	44.461	48.278	50.993	56.892
29	30.283	31.697	33.711	35.139	37.881	39.087	42.557	45.722	49.588	52.336	58.301
30	31.316	32.754	34.800	36.250	39.033	40.256	43.773	46.979	50.892	53.672	59.703
35	36.475	38.024	40.223	41.778	44.753	46.059	49.802	53.203	57.342	60.275	66.619
40	41.622	43.275	45.616	47.269	50.424	51.805	55.758	59.342	63.691	66.766	73.402
45	46.761	48.510	50.985	52.729	56.052	57.505	61.656	65.410	69.957	73.166	80.077
50	51.892	53.733	56.334	58.164	61.647	63.167	67.505	71.420	76.154	79.490	86.661
55	57.016	58.945	61.665	63.577	67.211	68.796	73.311	77.380	82.292	85.749	93.168
60	62.135	64.147	66.981	68.972	72.751	74.397	79.082	83.298	88.379	91.952	99.607

Statistics is a challenging subject. Add to this the challenge of computer coding and many would be ready to give up. In this text, Darrin Thomas explains basic concepts of statistics within the framework of using R. The blending of statistics and computer coding has quickly become a standard in research to in both academia and industry. As such, the concepts in this text are pertinent for the 21st century.

Su Jin So La

ISBN-13:978-1719554299
ISBN-10:1719554293

www.ingramcontent.com/pod-product-compliance
Lightning Source LLC
Chambersburg PA
CBHW062351220526

45472CB00008B/1768